Graphical Representation of Multivariate Data

Academic Press Rapid Manuscript Reproduction

Proceedings of the Symposium on
Graphical Representation of Multivariate Data
Naval Postgraduate School
Monterey, California
February 24, 1978

Graphical Representation of Multivariate Data

EDITED BY

PETER C. C. WANG

Naval Postgraduate School
Monterey, California

ACADEMIC PRESS **New York San Francisco London 1978**
A Subsidiary of Harcourt Brace Jovanovich, Publishers

ACADEMIC PRESS, INC.
111 Fifth Avenue, New York, New York 10003

United Kingdom Edition published by
ACADEMIC PRESS, INC. (LONDON) LTD.
24/28 Oval Road, London NW1 7DX

Library of Congress Cataloging in Publication Data

Symposium on Graphical Representation of Multivariate Data, Naval Postgraduate School, Monterey, Calif., 1978.
Graphical representation of multivariate data.

1. Multivariate analysis—Graphic methods—Congresses. 2. Multivariate analysis—Graphic methods—Data processing—Congresses. I. Wang, Peter C.C. II. Title.
QA278.S78 1978 519.5'3 78-10431
ISBN 0-12-734750-X

PRINTED IN THE UNITED STATES OF AMERICA

CONTENTS

LIST OF CONTRIBUTORS

Numbers in parentheses indicate the pages on which the authors' contributions begin.

James A. Ayers (183), Department of Mathematics, General Motors Corporation, General Motors Research Laboratories, Warren, Michigan 48090

William Black (199), Graduate School of Business Administration, University of Texas, Austin, Texas 78712

Bette Brindle (219), School of Information Studies, Syracuse University, Syracuse, New York 13210

Lawrence A. Bruckner (93), University of California, Los Alamos Scientific Laboratory, Los Alamos, New Mexico 87545

Lawrence Cahoon (243), U. S. Census Bureau, Suitland, Maryland 20233

Herman Chernoff (1), Department of Mathematics, Massachusetts Institute of Technology, Cambridge, Massachusetts 02139

Wayne W. Crouch (219), School of Information Studies, Syracuse University, Syracuse, New York 13210

Jerry K. Frye (219), Department of Speech Communication, State University of New York at Buffalo, Buffalo, New York 14214

Melvin J. Hinich (243), Department of Economics, Virginia Polytechnic Institute and State University, Blacksburg, Virginia 24061

David L. Huff (199), Graduate School of Business Administration, University of Texas, Austin, Texas 78712

Robert J. K. Jacob (143), Code 7509, Naval Research Laboratory, Washington, D.C. 20375

Gerald Edwin Lake (13), U. S. Navy, 5100 Kittery Landing, Virginia Beach, Virginia 23400

Gary C. McDonald (183), Department of Mathematics, General Motors Corporation, General Motors Research Laboratories, Warren, Michigan 48090

Juan E. Mezzich (123), Department of Psychiatry and Behavioral Sciences, Stanford University, Stanford, California 94305

Carol M. Newton (59), Department of Biomathematics, Center for Health Sciences, University of California, Los Angeles, California 94305

Peter C. Ordeshook (243), S.U.P.A., Carnegie-Mellon University, Pittsburgh, Pennsylvania 15213

David W. Scott (169), Department of Community Medicine, Baylor College of Medicine, Texas Medical Center, Houston, Texas 77030

Richard A. Tapia (169), Department of Mathematical Sciences, Rice University, Houston, Texas 77005

James R. Thompson (169), Department of Mathematical Sciences, Rice University, Houston, Texas 77005

Peter C. C. Wang (13), Departments of Mathematics and National Security Affairs, Naval Postgraduate School, Monterey, California 93940

David R. L. Worthington (123), Department of Psychiatry and Behavioral Sciences, Stanford University, Stanford, California 94305

PREFACE

This volume contains the Proceedings of the Symposium on Graphical Representation of Multivariate Data held at the Naval Postgraduate School in Monterey, California on February 24, 1978, with financial support from my research sponsors—Naval Intelligence Support Center, Naval Electronics Command, and the Office of Naval Research.

The principal reason for having this symposium was that there was no single source of literature for those interested in applications of graphical display of multivariate data from a wide range of different fields such as statistics, economics, regional planning, clinical research, social/political science, psychiatric studies, international relations, international trade, and arms transfer. The symposium brought together workers from diverse areas of interest to exchange ideas and views on utilization of graphical data displays. We hope that this publication of the papers presented at the symposium will be useful stimulation for others interested in data analysis.

The amount of scientific, technological, and policy information has been increasing and will continue to increase at an alarming rate. The need to develop techniques to analyze data problems of multivariate nature with high dimensions is eminent. The advent of powerful computer-based interactive graphic display systems now makes it possible to portray and analyze data in ways hitherto uneconomical. New interactive graphical systems are expected momentarily (see *Automated Production, Storage, Retrieval, and Display of Digitized Engineering Data*, Peter C. C. Wang, Western Periodicals, 1977), which will make it possible to graphically represent multivariate data still more economically. Realizing that we have the physical capability and the need to handle data graphically, we devoted ourselves at this symposium to the exploration and expansion of our knowledge beyond the conventional statistical techniques. And this expansion of knowledge into graphical representation of data is only the beginning of vastly more effective methods of communicating information.

My thanks go to all the authors whose papers appear here for their cooperation, which enabled me to produce this volume smoothly. Bud Goode's presentation on football forecasting was very interesting; I wish that he had had time to write the paper, but he did not. Finally, no list of credits would be complete without noting with appreciation the excellent help of Carolyn Quinn, Carol St. Onge, and Louise H. La Fosse, who prepared the final typing of the proceedings.

GRAPHICAL REPRESENTATIONS AS A DISCIPLINE

Herman Chernoff [1]

Department of Mathematics
Massachusetts Institute of Technology
Cambridge, Massachusetts

I. INTRODUCTION

Having spent the last few years on research removed from graphical representations, I welcome this symposium as an opportunity to review the field from a distance and to reflect on aspects that were less than perfectly clear a short time ago and are still somewhat hazy.

My first serious contact with this field occurred many years ago as an undergraduate student in an elementary statistics course where the students were obliged to present an india ink drawing of a histogram of data. My sloppy drawing was greeted with horror by the teaching assistant who insisted that it was unacceptable and had to be redone. That was bad enough but what really annoyed me was his reaction to my revised drawing. After studying the finished neat histogram which took hours of labor, he congratulated me, telling me that with a few drawings like this in my portfolio I would find it easy to gain employment as a statistician.

Considering the difficulty of finding jobs during those depression years, he evidently meant well. However, if the work of a statistician consisted of such non-intellectual labor, that profession was not for me.

Several years later I found myself drawn back to Statistics and eventually became a Professor of Statistics. My disrespect for drawing graphs remained and was reflected in part by the refusal to spend more than one lecture per course on histograms and ogives. Never never would I discuss pie charts nor diagrams with little men of various size. On the other hand, I found myself counseling the students over and over again to draw diagrams and graphs and to "look" at the data.

[1] Supported in part by ONR Contract N00014-75-C-055 (NR-042-331)

ISBN 0-12-734750-X

What seems to have been involved in my behavior is a combination of standard psychological reactions. The great success of some graphical techniques is based on their simplicity and transparency. While the innovation required substantial intellectual insight, college students have grown up frequently exposed to these simple techniques and should not need to have them explained. Indeed the college student who doesn't understand graph paper will probably not be helped much by explanations. Another reaction is that what is easily understood seems trivial. Thus the teacher tends to underestimate the conceptual difficulties in aspects of his subject which may be very novel to his students. In the process of explaining such sophisticated concepts as estimation and hypothesis testing, the teacher may forget that not everyone knows how graphs may be used for quick efficient calculations, nor that naive students may not conceive of learning what a formula means by drawing a graph. The more natural one finds the use of graphs, the less likely one is to feel that their usefulness requires explanation.

While everyone ought to know what a pie chart is, there are many intelligent people who do not tend naturally to think in terms of graphs and who find it relatively difficult to be innovative in the use of graphs and charts. For them it is of some importance to be exposed to a list of alternative methods of wide applicability, with detailed explanations of how to use these techniques as well as how not to use them.

For those more gifted with geometric insights, a few examples of innovative special purpose graphs can serve as an excellent introduction to the world of graphics. In any case there is no doubt that students of Statistics should be encouraged to use graphical techniques.

II. HISTORY

It has been widely noted, e.g. [6], that between 1940 and 1960 there had been a great decline in the attention paid by academic statisticians to graphical representation of data. Academic statisticians found the new analytical and conceptual aspects of their field more exciting. Lately, however, there has begun a substantial change in the attention paid to graphics as an intellectual discipline with important uses. This is reflected in part by several recent articles discussing Graphics and its history. Notable among these are those of Beniger and Robyn [2], Fienberg [6], Kruskal [9], and Royston [11].

One of the surprising facts arising from these articles is that the use of graphical techniques as an accepted practice dates back less than 200 years to the work of Crome and Playfair. Royston has suggested that Playfair, whose work was most influential, may have become acquainted with such techniques when he worked as a draftsman for Boulton and Watt.

It is possible that graphical techniques were commonly used by scientists and engineers before Playfair's work lead to their development in statistical applications.

One may wonder why these techniques were not used more often in the past. Granted that Statistics was a new discipline, why weren't such techniques used elsewhere and why were they used so rarely in deference to tables in the nineteenth century? It is possible that the technology of printing charts in books was insufficiently well developed to encourage the formal use of charts in publications in spite of their numerous advantages.

Whatever was the role of technology in the past, it is clear that the development of computers has had a considerable impact on the reawakening interest in graphics. The computer has the capacity to produce vast amounts of answers. Thus the computer presents the problem of digesting great amounts of data, a problem for which graphical methods are well suited.

A second problem presented by the computer is that it encourages one to feed in data and receive the output of stereotyped package programs too passively. When regressions were tediously carried out by hand, discrepant data and inappropriate models were apt to be discerned in the laborious process of computing. An answer spit out by a reliable computer is likely to carry conviction unless it is ridiculously wrong. Thus it is especially important to build diagnostic devices into the program. Here again graphical techniques such as plotting residuals are potentially valuable.

Not only does the computer provide the needs for graphical methods, it also furnishes a tool to help produce the graphs with minimal labor (and sometimes with insufficient thought).

II. CLASSIFICATION

In Fienberg's recent paper [6], he refers to Schmid's Handbook of Graphical Presentation [12] where suggestions for classifying charts and graphs are considered. Fienberg introduces an alternative classification of graphs in order to analyze the use of graphics in the Journal of the American Statistical Association (JASA) and Biometrika from 1920 to 1975. For a historical study of trends of use, some such classification may be required. However, this one leaves much to be desired as a guide for developing a discipline of graphics.

To quote two of Fienberg's references to Schmid:

"The qualities and values of charts and graphs as compared with textual and tabular forms of presentation have been succinctly summarized by Calvin Schmid (1954) in his <u>Handbook of Graphic Presentation</u>:

1. In comparison with other types of presentation, well-designed charts are more effective in creating interest and in appealing to the attention of the reader.

2. Visual relationships, as portrayed by charts and graphs, are more clearly grasped and more easily remembered.
3. The use of charts and graphs saves time, since the essential meaning of large masses of statistical data can be visualized at a glance.
4. Charts and graphs can provide a comprehensive picture of a problem that makes possible a more complete and better balanced understanding than could be derived from tabular or textual forms of presentation.
5. Charts and graphs can bring out hidden facts and relationships and can stimulate, as well as aid, analytical thinking and investigation."

"Schmid (1954) suggests that the basis for classifying charts and graphs must utilize one or more of the following criteria: (A) purpose, (B) circumstances of use, (C) type of comparison to be made, (D) form. Under purpose he lists the following three categories:

(1) Illustration,

(2) Analysis,

(3) Computation."

Then Fienberg presents an augmented list of six purposes developed by him and his graduate student Philip Chapman for the JASA - Biometrika study, the object of which was to determine whether the relative volume of usage of graphics had changed and whether there had been a shift in purpose to which graphs were being applied. Fienberg demonstrates a substantial decline in the use of graphics associated with data in these two statistical journals.

Fienberg's augmented list of purposes, presented with some explanation, are:

"I. Graphs depicting theoretical relationships, such as probability density functions, contours of multivariate densities, values of risk functions, contours of multivariate densities, values of risk functions for different estimators, and theoretical descriptions of graphical methods.

II. Computational graphs and charts, used as substitutes for tables --- e.g. Fox's charts, monograms, and especially charts with small detailed grid lines.

III. Non-numerical graphs and charts --- e.g. maps, certain skull diagrams, Venn diagrams, flow charts.

IV. Graphs intented to display data and results of analysis --- e.g. time series charts, histograms, results of Monte Carlo studies, scatter plots (even those with an accompanying regression line).

V. Plots and graphs with elements of both data display and analysis -- e.g. charts from older papers involving primitive forms of analysis; graphs of posterior distributions.

VI. Analytical graphs -- residual plots, half-normal and other probability plots where conclusions are drawn directly from graph, graphical methods of performing calculations, spectrum estimates from time series."

The development of a discipline of graphics seems to require a somewhat different approach in classification than that suggested by Fienberg. My feeling is that both graphical methods and potential applications should be measured on each of a list of relevant attributes. This list should contain the three purposes of Schmid and it is possible that a chart can serve 2 or 3 of these purposes simultaneously and that a particular application calls for more than one of these. The list should also contain the qualities and values mentioned by Schmid above. A list will be presented in Section 5 after a brief digression.

IV. REPRESENTATIONS OF MULTIVARIATE DATA

Before listing the attributes of charts, we describe briefly 4 methods of presenting multivariate data graphically. Let $\underline{x} = (x_1, x_2, \ldots, x_k)$ be a k-dimensional vector to be described. The method of profiles represents x by a bar chart with the i-th bar at a height determined by x . A common variation consists of using a polygonal line connecting the points whose coordinates are (i, x_i).

A second variation is the star in which the polygon is wrapped around a circle. To be more precise, the polygon connects the points whose polar coordinates are $\rho_i = x_i$ and $\theta_i = 2\ i/k$. Fienberg and others refer to Siegel, Goldwyn and Friedman [13] for using stars. I wish to thank R. J. K. Jacob for supporting my impression that this method had been used, at least informally, in the past by supplying a reference to Brinton [3, p.80] who refers to stars critically as a form of chart that should be banished to the scrap heap.

Andrews [1] introduced the representation using the Fourier Series Representation of $\underline{x}$ by plotting

$$f(\underline{x}) = x_1/\sqrt{2} + x_2 \sin t + x_3 \cos t + x_4 \sin 2t + \ldots$$

over the range $-\pi \leq t \leq \pi$

Finally, Chernoff [4] developed a computer program which draws a cartoon of a face determined by 18 parameters such as length of nose, curvature of the mouth, size of eyes, etc. If $k \leq 18$, one may adjoin to $\underline{x}$, 18-k specified numbers and use the resulting 18 component vector as the 18 parameters of a face to be drawn by the computer. The resulting face represents $\underline{x}$.

V. ATTRIBUTES

We list many attributes for consideration in classifying charts. Each graphical method has many of these attributes in varying degrees. Each application requires various of these attributes to a greater or lesser extent. The key to the successful use of graphics should involve a matching of method and application in terms of the extents of the attributes required by the application and how well the method supplies these attributes.

The first three items refer to the purposes to be served. These are the following:

1. Illustrate or communicate.

   Here the object is to communicate to the audience information which has typically been analyzed and understood.

2. Analyze or comprehend.

   If data have not been well understood it is often useful to find a representation which permits the analyst to develop an understanding of what conclusions may be drawn and what relations and regularities exist. Graphs designed for analysis may be used frequently in "practice" but frequently do not appear in publications once their purpose of clarification has been served.

3. Compute.

   Some charts or graphs provide a means of doing relatively accurate computing of many desired quantities. Nomograms are especially designed for this purpose.

In dealing with multivariate data the stars and faces are potentially useful for analysis because the representations may present a gestalt which recalls psychologically meaningful objects or ideas. The Fourier series and the profile seem to be less valuable in producing an emotional response. It is plausible to expect a circular version of the Fourier series to be more effective in analysis just as stars are better than profiles. However, the Fourier series representation has computational advantages. Squared distance in the vector space corresponds to squared integral and Andrews [1] has pointed out that good discriminant functions can be constructed from such graphs.

Faces can be used to communicate information to a limited extent <u>after some training.</u> Thus a wide smile can be made to correspond to good business or to a healthy patient and will be readily understood once the observer is given the <u>translation.</u> In this symposium Jacob [8] has described an innovative approach to using faces effectively for describing psychiatric patients. Goode [7] has developed a variation of faces consisting of a model of a football player whose measurements are determined by the qualities of a football team.

Other attributes of importance follow.

4. Impact.

   If an important fact is to be communicated, it should be presented in a forceful fashion. A time series can have considerable impact in showing sudden changes or long term trends.

5. Mnemonic Character.

When much information is to be communicated, it may be important to present some so that it will be well remembered. When complex data are to be analyzed the representation must be such that potentially important bits are not forgotten while observing others.

It is in this respect that the star is a great improvement over the profile for the profile makes it easy to confuse the index of the component which sticks out with the next one. The star improves in two respects. For small dimension $(k \leq 8)$, the direction of each vertex is easily remembered. The characteristic shapes that the star attains recall objects that help the memory. The latter explanation applies also to faces.

6. Attraction.

A chart addressed to a casual audience should attract attention. Playfair's charts were attractive aesthetically. The potential viewer may pass by an ugly or complicated looking chart.

7a. Accuracy (Precision).

For computing purposes the chart should enable one to make precise measurements easily.

7b. Accuracy (Lack of Distortion).

It is easy to draw charts which are precise but present a distorted impression. This is to be avoided in honest work designed to communicate.

8. Compactness.

For many purposes a table is a compact precise instrument for presenting knowledge to be used for reference. Sometimes a graph is even more compact and almost as precise.

9. Comprehensiveness.

Generally it is desirable that a representation be as comprehensive as possible. However, the attempt to communicate too much information may distract one from the important facts or may repel the casual audience.

When one has not yet decided what is and is not important, it is valuable to have as much information as can be comprehended. Faces seem to be good for this in multivariate data examples.

10. Self Explanatory.

The incentive to study a chart depends greatly on how much explanation is required to use it. Novel techniques face greater difficulties in being comprehended easily than do standard techniques. The self-explanatory quality of a method depends in part on the education of the viewer. A standard form such as a bar chart needs little explanation for most people. Brinton's criticism of stars [3] was based in part on his assessment of their difficulty in being understood. Incidentally his main interest in charts was as a tool for communication.

11. Time.

The speed in incorporating the information is one of the major advantages of many graphical techniques. Information to be used for reference may not demand speedy access.

12. Dimensionality.

The methods described in the preceeding section were designed for multidimensional data. Superficially it seems that the profile is extremely limited in the number of dimensions that can be handled effectively with it. I would estimate that in increasing order of ability are the Andrews Fourier Series, stars and faces. Each of these can be improved. The Andrews method, modified to wrap around a circle should improve its ability. The use of an inner and outer star should help the stars. A pair of faces effectively doubles the number of parameters covered and probably with little decrease in ability to detect changes or relations.

13. Theoretical vs. Data.

Fienberg's classification [6] involved the distinction between charts based on data and those representing theoretical concepts.

14. Contrast or Sensitivity.

For computing purposes the ability to detect slight differences is valuable. In analysis, the ability to detect outliers is important. These abilities require techniques which emphasize differences. Probability paper exploits the ability to detect non-linearities.

15. Ease of Application.

Pencil and paper techniques are desirable when doing analysis. A technique such as faces would be impossible to apply without a computer. Even with the computer, it is desirable to have a set up that makes it possible to implement quickly, easily and flexibly, whatever technique is chosen.

16. Audience.

The expected users should help determine the appropriate characteristics to be stressed in the representations to be used. It may make sense to classify the audience on several dimensions such as seriousness and training.

As research in graphics develops, it is to be expected that new attributes will be understood to be important while some of those listed above will diminish in importance. This is especially so since there is some interrelatedness among the items on the list.

VI. FUNDAMENTAL TOOLS

To make a discipline of graphical representations will require related research in technology and psychology plus innovation and the experimental testing of techniques.

The relevance of technology is certainly clear in the use of faces which would be impractical without a computer. It has been suggested earlier that the slow development and growth in use of graphical methods prior to and during the nineteenth century may have been due in part to the difficulty in publishing charts.

The relevance of psychology is even more basic. If we look at our list of attributes, we see that many of these involve perceptions and psychological reactions and hence the tools to evoke these must be tailored to our psychological tendencies to respond.

One fundamental tool is the use of models. A model is an effective device for communicating when it is possible for the user to make an easy one to one transformation from the properties of the model to the important characteristics of the real problem. The use of 3-dimensional models in construction probably dates back before writing. Picture writing and maps are examples. Goode's football player [7] is an adaptation of the idea of faces to create a model.

VII. ATTRIBUTES OF FACES

In the last few years, the use of faces has attracted considerable attention. A certain amount of experimentation has taken place, e.g. [5]. Applications have ranged from craters on the moon [10] to Soviet Policy in Sub-Saharan Africa [14]. What are the graphical attributes well served by faces? We shall go systematically through our list of attributes to see what one should expect.

First let me propose some conjectures about the underlying psychological phenomena which make faces potentially useful. I believe that we learn very early to study and react to real faces. Our library of responses to faces exhausts a huge part of our dictionary of emotions and ideas. We perceive the face as a gestalt and our built-in computer is quick to pick out the relevant information and to filter out the noise when looking at a limited number of faces.

Our ability to distinguish between very similar faces involves a mechanism where the brain converts the face to a mental caricature on which it operates. Hence the cartoon caricatures of faces which resemble our mental caricature will probably be more effective as a graphical representation then either more realistic drawings or freakish caricatures far removed from our mental ones.

Using these conjectures and possibly others which have not been stated, and some experience with the use of faces, let us go through our list of attributes.

Faces are of little use to illustrate or communicate unless the audience is specially trained in which case they can be of limited use. They are potentially very useful to analyze multivariate data to discern large complex relationships or to detect changes and similarities. They are almost useless for computation.

The emotional reactions of humans to faces triggers an impact. Our built-in library of reactions to faces enhances the mnemonic character of this representation.

They can not be relied on for accuracy and are subject to distortion, but they are comprehensive and self-explanatory and remarkably quick to analyze. The dimensionality is high and the ability to detect contrasts is excellent.

The ease of application depends on the software. It is potentially rather simple in an environment where data can easily be entered into the computer and interactive facilities can prompt the user.

Fienberg [6] carried out a comparison among Stars, Andrews Fourier Series and Faces on data involving 8 specimens of first lower premolar teeth from 9 animals. These were 3 humans, 4 gorillas and orangutangs, and 2 chimpanzees. These data were those used by Andrews [1] in his paper introducing his method. The data for each individual were the 8 principal components. In this study the faces suffered by the comparison.

However, the first 2 principal components carried almost all of the information. The fact that the method of principal components (basically a linear analysis) is so effective indicates that there is little to gain in analysis by using the more exploratory technique of faces or for the matter either of the other methods.

The use of the Andrews method is sensible for it permits one to calculate relevant distances and good linear disseminators. While the stars are effective in distinguishing the three groups of animals, a two-dimensional graph of the first two principal components is clearer and does better than any of these exotic techniques.

The effectiveness of the faces was dependent on the two features used for the first two principal components. I suspect that the faces would do relatively better if the three methods were applied directly to the original data. Even so, this is a problem when standard analysis is quite effective and none of these techniques is especially necessary.

VIII. SUMMARY

The use of faces seems to be growing in popularity. It gives me great pleasure to participate in this symposium, a large part of which is devoted to the use of faces. It is clear that it is only one of a host of new innovative graphical methods that are being developed. As more methods are devised to use our growing technological capacities to cope with the problems of data representations, we must allocate some of our energy to develop graphics as a discipline.

This will require many inputs some of which are already flowing in greater quantities. In addition to innovative ideas, we need to develop further the technology of interactive statistical computing, we need to understand the psychology of perception better, and we have much serious experimenting to do with graphical techniques. I hope that the classification concept proposed here will help clarify the issues to be settled by experimentation and help also in the proper design of these experiments.

REFERENCES

1. Andrews, D. F. (1972). "Plots of High Dimensional Data," Biometrics, 28:125.
2. Beniger, James R. and Robyn, Dorothy L. (1978). "Quantitative Graphics in Statistics: A Brief History," American Statistician, 32:1.
3. Brinton, W. C. (1923). Graphic Methods for Presenting Facts, The Engineering Magazine Company, New York.
4. Chernoff, Herman (1973). "Using Faces to Represent Points in K-dimensional Space Graphically," Journal of the American Statistical Association, 68:361.
5. Chernoff, Herman and Rizvi, M. Haseeb (1975). "Effect on Classification Error of Random Permutations of Features in Representing Multivariate Data by Faces," Journal of the American Statistical Association, 70:548.
6. Fienberg, S. E. (1977). Graphical Methods in Statistics, Tech. Report No. 304, Dept. of Applied Statistics, University of Minnesota, pp. 44.
7. Goode, B. (1978). "Current Ways of Applying Graphical Representation of Multivariate Data to Football Forecasting," Proc. of Symposium on Graphical Representation of Multivariate Data (to be published).
8. Jacob, R. J. K. (1978). "Facial Representation of Multivariate Data," Proc. of Symposium on Graphical Representation of Multivariate Data (to be published).
9. Kruskal, William (1977). "Visions of Maps and Graphs." Proceedings of the International Symposium on Computer-assisted Cartography, 1975, pp. 27.
10. Pike, J. (1974). "Craters on Earth, Moon and Mars: Multivariate Classification and Mode of Origin," Earth and Planetary Science Letters, 22:245.
11. Royston, E. (1970). "Studies in the History of Probability and Statistics, III. A Note on the History of the Graphical Presentation of Data," Biometrika, Pts. 3 and 4 (December 1956), pp. 241; reprinted in Studies in the History of Statistics and Probability Theory, eds. E. S. Pearson and M. G. Kendall, London: Griffin.
12. Schmid, Calvin F. (1954). Handbook of Graphic Presentation. Ronald Press, New York.
13. Siegel, J. H., Goldwyn, R. M. and Friedman, H. P. (1971). "Pattern and Process of the Evolution of Human Septic Shock." Surgery, 70:232.
14. Wang, P. C. C. and Lake, G. E. (1977). "A Graphical Representation of Multivariate Data: Development and Application to an Assessment of the Soviet Influences in Sub-Saharan Africa." Proc. of Symposium on Graphical Representation of Multivariate Data (to be published).

APPLICATION OF GRAPHICAL MULTIVARIATE TECHNIQUES IN POLICY SCIENCES

Peter C. C. Wang[1]

Departments of Mathematics and National Security Affairs
Monterey, California

LT Gerald Edwin Lake, USN[2]

ABSTRACT

This paper is an application of the graphical representation of k-dimensional data technique developed by Professor Herman Chernoff. It is written for the intended use of foreign policy and international relations analysts as a statistical tool in analyzing and presenting complex Soviet foreign policy phenomena. It discusses building a computer base data file from existing sources, purpose and techniques for data modifications and transformations, and methods for selecting variables for the research problem. Soviet foreign policy acts in twenty-five Sub-Saharan African countries for 1964 through 1975 are then represented and analyzed in ten sets of FACES. Other applications of the FACES methodology are reviewed and recommendations are made for further modifications and applications of the methodology in the area of foreign policy studies.

I. INTRODUCTION

The object of this paper is to represent multivariate data graphically. Using graphical means for communicating or explaining phenomena is as old as man himself, as the following examples illustrate: 1) Every person who has traveled in a foreign country where he could not communicate in the native language has probably resorted to drawing pictures to express his ideas or needs. 2) The pictures that have been discovered in caves and tombs have revealed numerous facts about the lives and social conditions of their creators. 3) If you have ever gone to a carpenter or machinist to have something built and have given him a complete set of instructions, nine times out of ten he will say, "Draw me a picture." In summary, people are accustomed to communicating with graphics.

---

[1] Research is partially sponsored by NISC (Naval Intelligence Support Center and NAVLEX (Naval Electronics Long Range Planning Group).

[2] Present address: 5100 Kittery Landing, Virginia Beach, Virginia 23400.

ISBN 0-12-734750-X

The graphics program selected for this research is the face drawing program of H. Chernoff. The graphic representation of this method is a face with eyes, nose, mouth, ears and eyebrows which change when the variables which are being measured change. This is done by representing a point in k-dimensional space, (k is a scalar which defines the number of points) and changing its position according to changes in variables.

The purpose for graphical representation of multivariate data is to enhance the ability of the analyst to detect and comprehend important phenomena, serve as a mnemonic device for remembering major conclusions, communicate major conclusions to others, and provide a means for doing complex and relatively accurate calculations informally. (Chernoff, 1971) The use of a face enhances the utility of graphics techniques in accomplishing these goals by drawing on the observer's culturalization process that teaches him to recognize and respond to changes in facial expressions. It assumes that the human eye functions more like an analog computer than like a digital computer. An analog computer uses continuous input whereas a digital computer reads, records and reads again in a step by step procedure. The analyst when analyzing the data is looking for clusters and not even distribution. It is precisely the clustering that lends graphics its power as an analytic tool. In the face drawing program, multivariate data that have similar properties will also be represented by similar faces. These similar clusters are called, for simplicity, families.

The subject of this analysis is Soviet foreign policy in Sub-Saharan Africa. Information has been collected which measures amounts, levels, and frequencies of Soviet activities with lesser developed countries (LDC's). Section II presents the method of data selection, examination and storage on a quick access device. Necessary modifications and transformations of the data are also discussed.

The data file was still large and some decisions had to be made about how to reduce the number of variables, without removing any important ones. Because of this last criterion, factor analysis techniques were not used. Factor analysis would have indicated, for example, that this group of variables is highly correlated, and only one of the variables or their combined factor score should be used. What is needed is a system or methodology that is more concerned with the quality of the relationship and not the quantity. The scalogram and even analysis techniques discussed in Section III are believed to have the property of measuring the qualitativeness as well as the quantitativeness of the relationships.

Sections IV, V and VI are concerned with the face drawing program. A step by step procedure is presented in Section IV of the methodology for making the necessary changes to the computer program in order to draw faces. Section V presents the results of the faces that were computed for twenty-five selected countries of Sub-Saharan Africa, and Section VI presents a discussion of some of the faces which are being computed at other institutions.

A. Sources of Data

1. Orchestration Project File. The Orchestration File is a computer adaption of Aid and Trade Activities of Communist Countries in Less Developed Areas of the Free World. The unit of analysis is the nation state, and the interval is the calendar year.

The file copy received from CNA was transferred to a 2311 disk pack and stored in SPSS (Nie et al., 1975) format in five separate files to reduce access time. The five files are: WORLD, AFRICA, ASIA, ARAB, and LATIN AMERICA. The files include data for twenty-one years (1954 to 1975) and ninety-one variables. The variables are organized in the following manner:

1. Variables one through ten (VAR001 - VAR010) are descriptive variables that identify the country by name and code and indicate the date of independence and the date of admission to the United Nations; only post-independence data is included for each country.

2. Variables eleven through fifty-six (VAR011 - VAR056) and seventy-seven through eighty-two (VAR077 - VAR082) and eighty-six through ninety-one (VAR086 - VAR091) describe various Soviet, East European and People's Republic of China's interactions with the 111 LDC's. See Table 1 for a complete breakdown.

3. Variables fifty-seven through seventy-six (VAR057 - VAR076) are dichotomous variables which indicate whether or not that country has the property described. See Table 2 for a breakdown of these variables.

4. Variables eighty-three through eighty-five (VAR083 - VAR085) are country descriptive variables (CNA, 1976). They are:

VAR083 - Yearly Defense Expenditures
VAR084 - Gross National Product
VAR085 - Men in Armed Forces in Thousands

2. WARP File. WARP was designed for use by the Directorate of Plans and Evaluation Special Regional Studies Division in the Office of the Secretary of Defense. (Christensen and Pieper, 1976) Its purpose is to enable defense analysts to make various assessments of U.S. worldwide interests and associated threats on a quantitative basis. The data base includes country, regional, and worldwide variables which are related to the demographic, political, socio-economic, military, and commercial characteristics of the defense-related areas. (See Appendix A for a list of the 195 Countries.)

There are four major categories of WARP Data: 1) General, 2) Money, 3) People, and 4) Trade. The data base is divided into these categories both logically (for access and use) and physically (for computer organization and storage). Appendix A includes a flow chart of the interrelationship of the WARP files.

TABLE I

BREAKDOWN OF THE ORCHESTRATION FILE

VARIABLE NAME	SOVIET	EAST EUROPEAN	PRC
Airline flights to LDC	VAR011	VAR012	
Number of economic aid agreements	VAR013	VAR014	VAR015
Number of military aid agreements	VAR016	VAR017	VAR018
Total value of military aid agreements	VAR019	VAR020	VAR021
Economic aid extensions	VAR022	VAR023	VAR024
Military aid grants and discounts	VAR025	VAR026	VAR027
Total economic aid drawings to date	VAR028	VAR029	VAR030
Yearly economic aid drawings	VAR031	VAR032	VAR033
Military personnel in LDC	VAR034	VAR034	VAR035
LDC military departing for training in the ...	VAR036	VAR037	VAR038
LDC military personnel training in the ... in December	VAR039	VAR040	VAR041
Economic aid technicians in the LDC	VAR042	VAR043	VAR044
Technical traines departing for training in the ...	VAR045	VAR046	VAR047
Technical trainees training in the ... in December	VAR048	VAR049	VAR050
Academic students departing for training in the ...	VAR051	VAR052	VAR053
Academic students training in the ... in December	VAR054	VAR055	VAR056
Imports from LDC	VAR079	VAR078	VAR077
Exports to LDC	VAR082	VAR081	VAR080
Merchant ship calls	VAR086	VAR087	VAR088
Arms deliveries	VAR089	VAR090	VAR091

TABLE II

TWENTY DICHOTOMOUS VARIABLES IN THE ORCHESTRATION FILE

VARIABLE NUMBER	VARIABLE NAME
VAR057	Offshore Hydrocarbon Producers
VAR058	Major Oil Producers
VAR059	Major Mineral Producers
VAR060	Distant Fishing States
VAR061	Major Fishing State Nominal Catch
VAR062	Landlocked States
VAR063	Shelf-locked States
VAR064	Narrow Shelf States
VAR065	Broad Shelf States
VAR066	Strait States
VAR067	Blue Water Navy States
VAR068	Coastal Navy States
VAR069	African States
VAR070	Asian States
VAR071	Latin American States
VAR072	Arab States
VAR073	Narrow Straits States
VAR074	Oceanic Archipelago States
VAR075	Coastal Archipelago and Distant Archipelago States
VAR076	States on Semi-enclosed Seas

General. This category is composed of all variables which do not fit in the other categories. The bulk of GENERAL was extracted from the Cross-National Time-Series Data Archive of the Center for Comparative Political Research (CCPR), State University of New York. There are diverse social, political, and demographic characteristics collected for each of the 167 independent countries. In addition to the CCPR data, counts of treaties between countries, number of naval port calls, number of national banks, and United Nations voting consensus data are included. There are currently 69 data variables in GENERAL.

Money. Data related to major power interests abroad is shown by various monetary factors. Major data types include bilateral trade amounts, civil and military assistance programs, and products. There are currently 47 data variables in MONEY.

People. This category contains a variety of information related to the number of military, civilian, and diplomatic personnel in foreign countries, foreign country population, number of immigrants, and other political-diplomatic variables descriptive of major power relations with foreign countries. There are currently 43 variables in PEOPLE.

Trade. This category contains descriptive information related to essential U.S. Foreign trade routes, and tonnage and value of commodities being traded over these international pathways. This category currently contains data related to a number of commodity types and 65 different trade routes.

Sub-Categories. Within each of the major categories, the WARP Data Base is sub-categorized to indicate the relationship of six major powers to 195 target countries. These relationships apply only to the GENERAL, MONEY and PEOPLE categories since the TRADE category represents only U.S. involvement. The data is organized in country-pair relationships.

B. SUB-SAHARAN AFRICA FILE

The file used to construct the faces of Soviet foreign policy is the Sub-Saharan Africa File. It was developed from the Orchestration and WARP files and consists of the twenty-five African countries shown in Figure 1. Like the Orchestration File, the unit of analysis is the nation state, and the interval is the calendar year. The period of the file is from 1963 through 1975. To be a candidate, a country had to be in Sub-Saharan Africa, independent by 1963, and underdeveloped. This file consists of 40 variables which are listed in Table 3.

1. Missing Values for Variables. The Orchestration File assigns a value of 99999 to missing entries. The WARP file only enters data for the years that complete data are available (Gurr, 1972). Therefore, when merging the two files, careful attention to each variable had to be given. For ease in handling, the method used in the Orchestration File

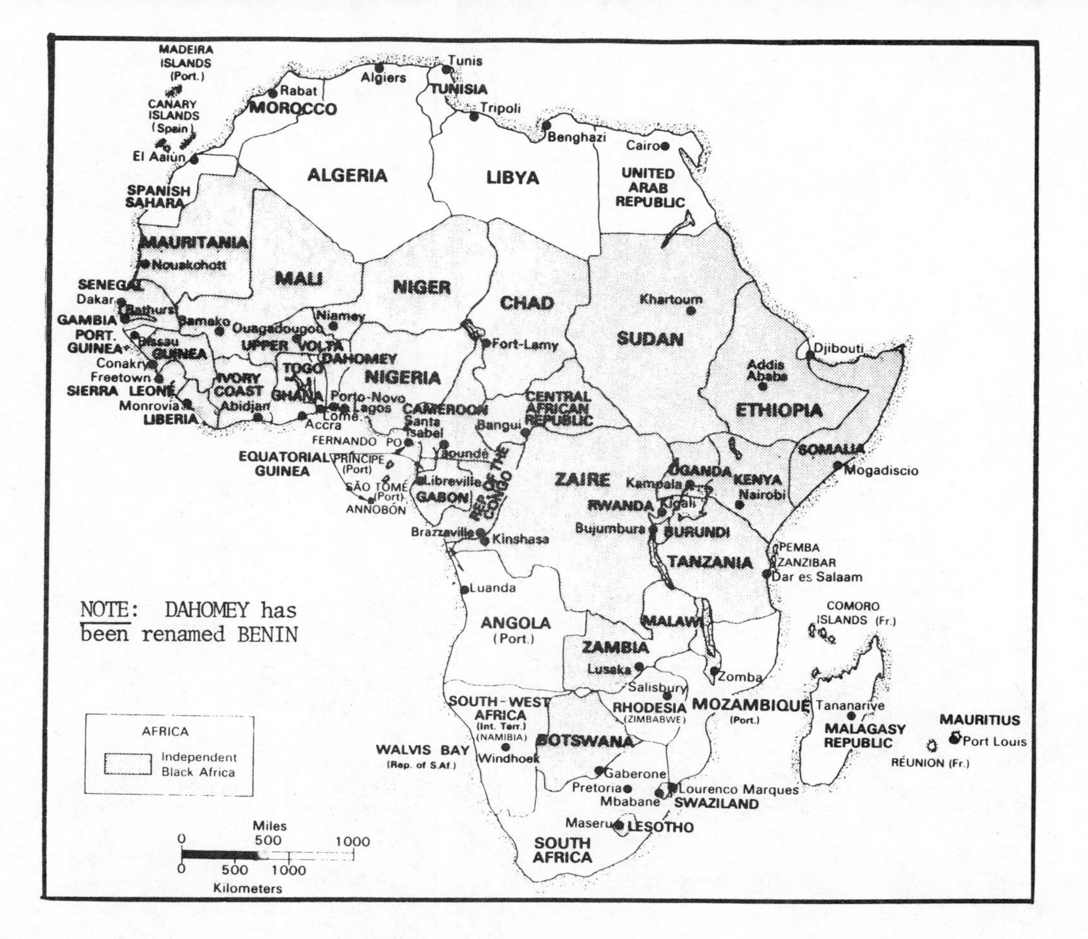

Figure 1. Map of Africa.

TABLE III

THE 40 VARIABLES IN THE SUB-SAHARAN AFRICA FILE

Variable Number	Description
V01	Year of the entry
V02	CNA country code
V03	Banks country code
V04	World Handbook country code
V05	Case ID (year times 1,000) plus CNA country code
V06	Soviet airline flights
V07	Number of economic aid agreements
V08	Number of military aid agreements
V09	Total value of military aid agreements
V10	Value of economic aid extended
V11	Value of military aid grant or discount
V12	Total economic aid drawn
V13	Yearly economic aid drawn
V14	Soviet and East European military personnel in the LDC
V15	LDC military personnel departing for training in the USSR
V16	LDC military personnel training in the USSR as of Dec. 1976
V17	Soviet economic technicians in the LDC
V18	LDC technicians departing for training in the USSR
V19	LDC technicians training in the USSR in Dec. 1976
V20	LDC academic students departing for school in the USSR
V21	LDC academic students in the USSR in Dec. 1976
V22	Imports from the USSR
V23	Exports to the USSR
V24	Size of the LDC defense budget
V25	LDC gross national product
V26	Size of the LDC armed forces
V27	Soviet merchant ship port visits
V28	Total value of soviet arms deliveries
V29	Total number of economic aid agreements
V30	Total number of military aid agreements
V31	Submarine port visits
V32	Major combatant port visits
V33	Minesweeper port visits
V34	Amphibious port visits
V35	Auxillary port visits
V36	Total number of Soviet naval port visits
V37	Size of the LDC defense budget
V38	LDC exports to the USSR
V39	LDC imports from the USSR
V40	Soviet arms deliveries

was initially duplicated in the Sub-Saharan African File (99999 for missing values). However, missing values will produce erroneous results in the present FACES program. Four methods were selected to provide solutions for missing values and are presented in their precedence order (Lindsey, 1976).

a. If the variable selected has missing values and is recorded in both files, and one file has an entry for the missing value, and the entries which are present correlate, then the entry from the other file is to be read into the missing value entry.

b. If the missing value is an end point and does not satisfy the first criteria, then an exponential smoothing method is to be used to compute the missing value. The advantage of exponential smoothing is that it assumes the adjacent value is the best indicator as to the magnitude of change of the missing value. It computes the geometric mean and a scalar multiplier which is most influenced by the adjacent variable.

c. If one data point is missing and it is not an end point, compute the missing value by averaging the entry before and after.

d. If more than one data point is missing, compute the missing values by the least squares technique.

2. Data Transformation. Some modification (normalization) of the data is necessary to produce the best results (Gurr, 1972). To use the data as they were collected might produce misleading results. For example, a maximum number of port visits by any country for any year is 60 and the mean value is 3.1. Observation of the data in its present form could be an indication that it is not very difficult to be an average country because such a country has to receive only three port visits per year. In reality it is the first few port visits that are the most difficult to achieve. What is required to properly analyze port visits is the mathematical property of port visits that will put the mean in the center of the distribution and the extremes equal opposite distances from the mean. Therefore, it can be said that the distribution of port visits is not normal. A normal function would produce a bell-shaped curve when all the data points are plotted, with mean in the center of the curve. Figures 2 and 3 illustrate these distributions.

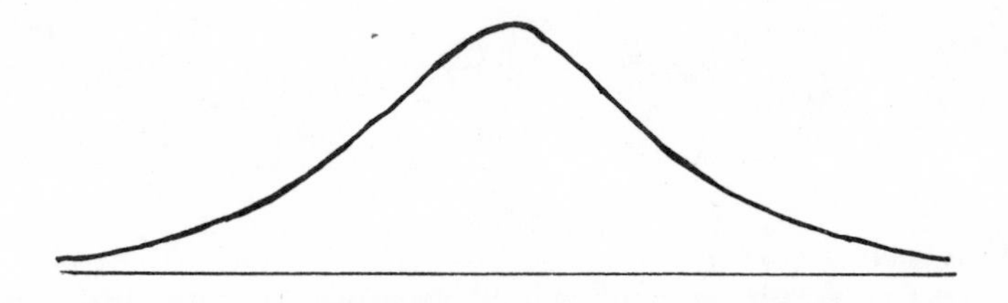

Figure 2. Normal distribution curve.

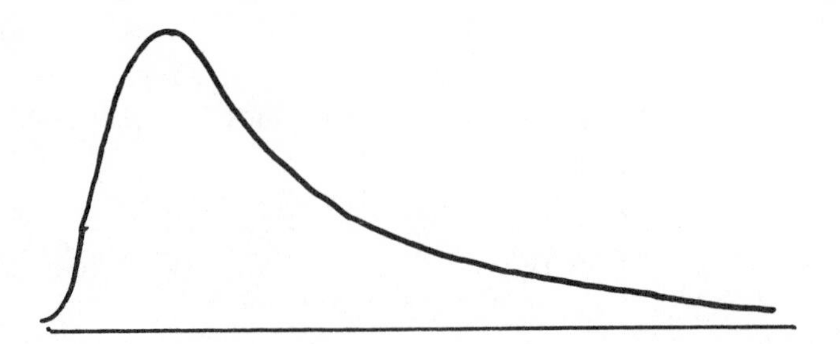

Figure 3. Distribution curve for port visits.

Another and more readily available indication of distribution other than normal is found in examining the skewness and kurtosis (Nie *et al.,* 1975).[2] Skewness measures deviations from symmetry. It is the third moment and will take the value of zero when the distribution is a completely symmetric bell-shaped curve. A value less than +3 and greater than -3 is desirable. Positive values indicate the data is clustered more to the left of the mean with the extreme values off to the right. Negative values indicate just the opposite condition. In the port visits example, the skewness is 4.395.

Kurtosis is the measure of the peakedness or flatness of the curve. A value of zero indicates normal distribution. Positive values indicate the distribution is peaked with negative values indicating flatness. A value less than +5 and greater than -5 is desired. Kurtosis is the fourth moment. The kurtosis for the example is 20.560.

Experience and/or intuition tells the analyst that port visits are important and just because the data have a skewness and kurtosis that indicate peakedness and uneven distribution does not imply that visits are not important. But to use the data statistically, one should examine them to determine if they have a property that is normally distributed. What the analyst would find particularly useful in this case is a difficulty factor, e.g., how difficult it is for the Soviets to penetrate an African country in the form of port visits. The difficulty factor would be normally distributed and a function of port visits. For this new

[2] Equation for skewness:

$$\frac{\sum_{i=1}^{N} \left((X_i - \overline{X})/s \right)^4}{N}$$

Equation for kurtosis (centralized):

$$\frac{\sum_{i=1}^{N} \left((X_i - \overline{X})/s \right)^4}{N}$$

variable, difficulty factor for port visits, it would then be as difficult to move from no port visits to the average as it would be to move from the average to the maximum. Plotting the values for the new variable, difficulty factor for port visits, would produce a bell-shaped curve. For port visits the difficulty factor is assumed to be equal to the fourth root of the number of port visits for a country per year. Taking the fourth root for each element in that variable would produce a transformed variable that is approximately normally distributed with a skewness of 2.839 and a kurtosis of 0.411. Another explanation of the same data would be: a country moving from two port visits to three is much more significant an indicator of Soviet penetration than another country moving from forty-four to forty-five. A change from two to three is as significant as a change from forty-four to fifty-seven in the untransformed data.

Table 4 is a breakdown of the seventeen variables (indicating their skewness and kurtosis) used in the FACES program and, when appropriate, their transformed values.

III. SELECTING THE VARIABLES

Three methods of analysis were used for exploratory research to select the variable to be used in the FACES program. The three methods were Guttman scaling, WEIS (World Event/Interaction Survey) system, and trend analysis. They will be presented in the next three sections along with their results.

A. Guttman Scaling

Guttman scaling is a statistical technique which examines the underlying operating patterns which possibly connect three or more items (in this study, ten items) to determine if the interrelationships of those items meet several special properties which define a Guttman scale (Nie _et al._, 1975 and Gurr, 1972). The two most important properties are unidimensionality and cumulativeness. Unidimensionality requires that the variables must measure movement towards or away from the same single objective. In this study, the objective is Soviet penetration into Africa. Cumulativeness implies that the variable can be scaled by the degree of difficulty that a LDC has in achieving each variable. In this study, cumulativeness implies how difficult it was for the Soviets to establish themselves in Africa through certain specific activities, such as placing Soviet military personnel in the country.

To analyze the results and to determine whether the scale is valid, four statistics are produced, i.e., the coefficient of reproducibility (C_R), the minimum marginal reproducibility (M_{MR}), the percent improvement ($I_{\%}$), and the coefficient of scalability (C_S). The two most important statistics and the minimum values required to have a valid scale are $C_R = 0.90$, $C_X = 0.60$. Figure 4 shows the results of scaling ten variables and will be used in the explanation of the statistics.

TABLE IV

CONDESCRIPTIVES

Variable	Name	Kurtosis	Skewness	Function	Adjusted Kurtosis	Adjusted Skewness
V09	Military Aid Extended	42.586	6.313	$\sqrt[4]{V09}$	10.467	3.343
V10	Economic Aid Extended	76.824	8.107	$\sqrt[4]{V10}$	7.013	2.734
V11	Military Grants and Discounts	85.208	8.978	$\sqrt[4]{V11}$	19.847	4.442
V13	Yearly Economic Aid Drawn	86.898	8.054	$\sqrt[4]{V13}$	-0.276	0.888
V14	Soviet Military in LDC	51.706	6.602	$\sqrt[4]{V14}$	1.024	1.488
V15	LDC Military Training in USSR	30.035	5.093	$\sqrt[4]{V15}$	3.271	2.166
V17	Soviet Economic Technicians in LDC	10.379	2.982	$\sqrt[4]{V17}$	-1.133	0.388
V18	LDC Technicians Training in USSR	42.453	6.095	$\sqrt[4]{V18}$	4.036	2.409
V21	LDC Academic Students in USSR	1.606	1.362	V21	1.606	1.362
V22	LDC Imports from USSR	32.808	5.052	$\sqrt[4]{V22}$	-0.437	0.703
V23	LDC Exports to USSR	13.965	3.351	$\sqrt{V23}$	1.787	1.427
V24	Size LDC Defense Budget	38.768	6.003	$\sqrt[4]{V24}$	4.864	1.858
V25	Size LDC Gross National Product	18.245	3.802	$\sqrt{V25}$	4.206	1.767
V26	Size LDC Armed Forces	35.935	5.735	$\sqrt[4]{V26}$	3.997	1.669
V29	Total number Economic Aid Agreements	4.675	2.009	V29	4.675	2.009
V30	Total number Military Aid Agreements	5.169	2.311	V30	5.169	2.311
V36	Total number Soviet Naval Port Visits	20.560	4.395	$\sqrt[4]{V36}$	6.411	1.263

ITEM..	V11		V09		V10		V18		V15		V14		V30		V29		V17		V21		
RESP..	0	1	0	1	0	1	0	1	0	1	0	1	0	1	0	1	0	1	0	1	TOTAL
SCALE 10	0	1 ERR	0	1	0	1	0	1	0	1	0	1	0	1	0	1	0	1	0	1	1
9	0	1	1	0 ERR	0	1	0	1	0	1	0	1	0	1	0	1	0	1	0	1	1
8	2	0	2	0	0	2 ERR	0	2	0	2	0	2	0	2	0	2	0	2	0	2	2
7	14	1	13	2	13	2	4	11 ERR	0	15	0	15	0	15	0	15	0	15	1	14	15
6	39	0	37	2	36	3	26	13	11	28 ERR	2	37	0	39	4	35	1	38	0	39	39
5	23	0	23	0	23	0	20	3	23	0	3	20 ERR	0	23	0	23	0	23	0	23	23
4	10	0	10	0	9	1	9	1	10	0	9	1	2	8 ERR	1	9	0	10	0	10	10
3	68	0	68	0	68	0	64	4	68	0	67	1	60	8	11	57 ERR	2	66	0	68	68
2	35	0	35	0	35	0	33	2	35	0	35	0	33	2	19	16	20	15 ERR	0	35	35
1	55	0	55	0	55	0	55	0	55	0	55	0	55	0	55	0	55	0	0	55 ERR	55
0	23	0	23	0	23	0	23	0	23	0	23	0	23	0	23	0	23	0	23	0	23
SUMS	269	3	267	5	262	10	234	38	225	47	194	78	173	99	113	159	101	171	24	248	272
PCTS	99	1	98	2	96	4	86	14	83	17	71	29	64	36	42	58	37	63	9	91	
ERRORS	0	2	1	4	0	6	4	23	11	0	5	2	2	10	16	16	23	0	1	0	126

325 CASES WERE PROCESSED
53 (OR 16.3 PCT) WERE MISSING

STATISTICS..

COEFFICIENT OF REPRODUCIBILITY = 0.9537
MINIMUM MARGINAL REPRODUCIBILITY = 0.8096
PERCENT IMPROVEMENT = 0.1441
COEFFICIENT OF SCALABILITY = 0.7568

Figure 4. Guttman Scale of Sub-Saharan Africa 1963 through 1975.

A case consists of ten variables for one country for one year. In Figure 4 the figures in the right column labeled "Total" indicate the number of cases that passed the test to achieve the scale value listed in the left hand column labeled "Scale." The values in the variable column indicate the number of cases that passed and failed each item for the scale value indicated. The variable which was the most difficult for the Soviets to achieve is the variable placed in the column farthest to the left; the degree of difficulty decreases as one reads from left to right. The country which was most deeply penetrated by the Soviets is ten on the scale. The amount of Soviet penetration decreases as one reads from the top to the bottom of the scale.

The coefficient of reproducibility is equal to one minus the total number of errors (an error is an incorrect response for an item for a particular scale value), divided by the total number of responses:

$$
\begin{aligned}
C_R &= 1 - (\text{total errors}/\text{total responses}) \\
&= 1 - (126/2720) \\
&= 0.9537
\end{aligned}
$$

The minimum marginal reproducibility is the minimum coefficient of reproducibility that could have occurred for the scale given the division points used and the number of responses passing and failing each item. It is equal to the sum of the maximum marginals for each variable divided by the total number of respondents. In Figure 4 at the bottom of the matrix are the values "SUMS," "PCTS" and "ERRORS." In the row labeled "SUMS" are two values for each variable; the larger of the two is the maximum marginal for that variable:

$$
\begin{aligned}
M_{MR} &= (269 + 267 + 262 + 234 + 225 + 194 + \\
&\quad 173 + 159 + 171 + 248) \;/\; 2720 \\
&= 2202 \;/\; 2720 \\
&= 0.8096
\end{aligned}
$$

The percent improvement is the difference between the coefficient of reproducibility and the minimum marginal reproducibility:

$$
\begin{aligned}
I_{\%} &= C_R - M_{MR} \\
&= 0.9573 - 0.8096 \\
&= 0.1441
\end{aligned}
$$

The final statistic is the coefficient of scalability. It is equal to the percent improvement divided by one minus the minimum marginal reproducibility:

$$C_S = I_{\%} / (1 - M_{MR})$$

$$= 0.1441 / (1 - 0.8096)$$

$$= 0.7568$$

The results of scaling the ten variables produce a pattern of Soviet relationships with the twenty-five countries in the Sub-Saharan File. Eight of these variables were checked against the files on Africa and World with similar results. As a result of the scaling, the first ten variables were selected for the FACES program (see TAble 5 for these variables). The WEIS system provided the next variable.

Scale Value	Variable	Variable Name
10	V11	Soviet Military Grant or Discount
9	V09	Total Value of Military Aid Agreement
8	V10	Soviet Economic Aid Extended
7	V18	LDC Economic Technicians Training in the USSR
6	V15	LDC Military Training in USSR
5	V14	Soviet Military Personnel in the LDC
4	V30	Military Aid Agreements
3	V29	Economic Aid Agreement
2	V17	Soviet Economic Technicians in LDC
1	V21	Academic Students in USSR

Table V. The 10 Guttman scaled variables.

B. Weis System

The WEIS System is a machine readable coded content analysis of the New York Times. Each entry includes the actor (the country directing the activity), the target (country at which the activity is directed), and arena (area in which the event takes place), and an event code (ranks the activities ordinally from one to twenty-two with sub-categories). The entry also contains a brief summary of the article (Sherwin, 1973).

An events search of the WEIS System indicated a pattern of Soviet foreign policy activities, including port visits by Soviet Naval units, in three Sub-Saharan African countries. These results were similar to the results obtained in the Guttman scaling. Because the three patterns were similar, further tests were made applying Guttman scaling to port visits. The results indicated that port visits could be scaled and consequently, port visits were included in the list of variables for the FACES program.

The number of events coded for Africa is very low and not considered to be a reflection of actual activity in Africa. Instead the low number of events probably reflects a lack of reporting on Africa by the New York Times, or reflects the selection criteria of the person doing the coding. Although the WEIS system did not contain enough information to be used statistically for this project, it did serve as an indicator of some of the more important events, and it could effectively be used to search for patterns in other areas of the world.

The six remaining variables were chosen through trend analysis.

C. Trend Analysis

Figures 5 and 6 are graphs of Soviet imports and exports with the world's 111 LDC's and the 42 LDC's in the Africa File. Figure 5 indicates that in trade with the LDC's, worldwide, the balance of trade is in favor of the Soviet Union. But, Figure 6 reveals that in 1959, 1960, 1963, and 1965 the balance of trade between the Soviets and Africa was in favor of the African states.

Another interesting facet of Soviet trade with all LDC's (Figure 5) is the indication that from 1958 to 1964 there was a stable trading pattern. From 1964 to 1969 there appeared a noticeable increase in the balance of trade in favor of the Soviet Union, but indications in 1969 were that there was a decrease in Soviet exports paralleled by an increase in imports. The 1970 through 1974 period reflected a sharp increase in both imports and exports by the Soviet Union.

Another variable which indicates an interesting trend is the Total Value of Economic Aid Extensions. Figure 7 shows a steady increase in the amount of aid extended. This steady increase probably indicates that economic aid is planned and not an opportunistic variable.

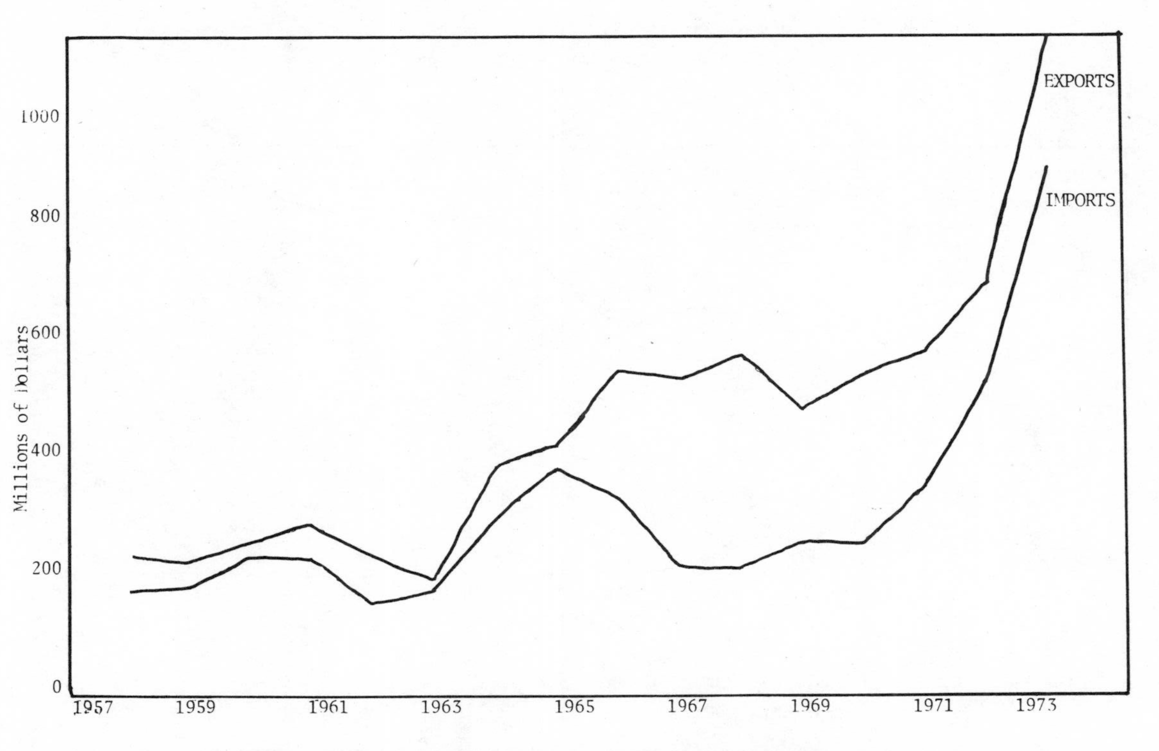

Figure 5. Soviet trade with 111 lesser developed countries 1958 through 1973.

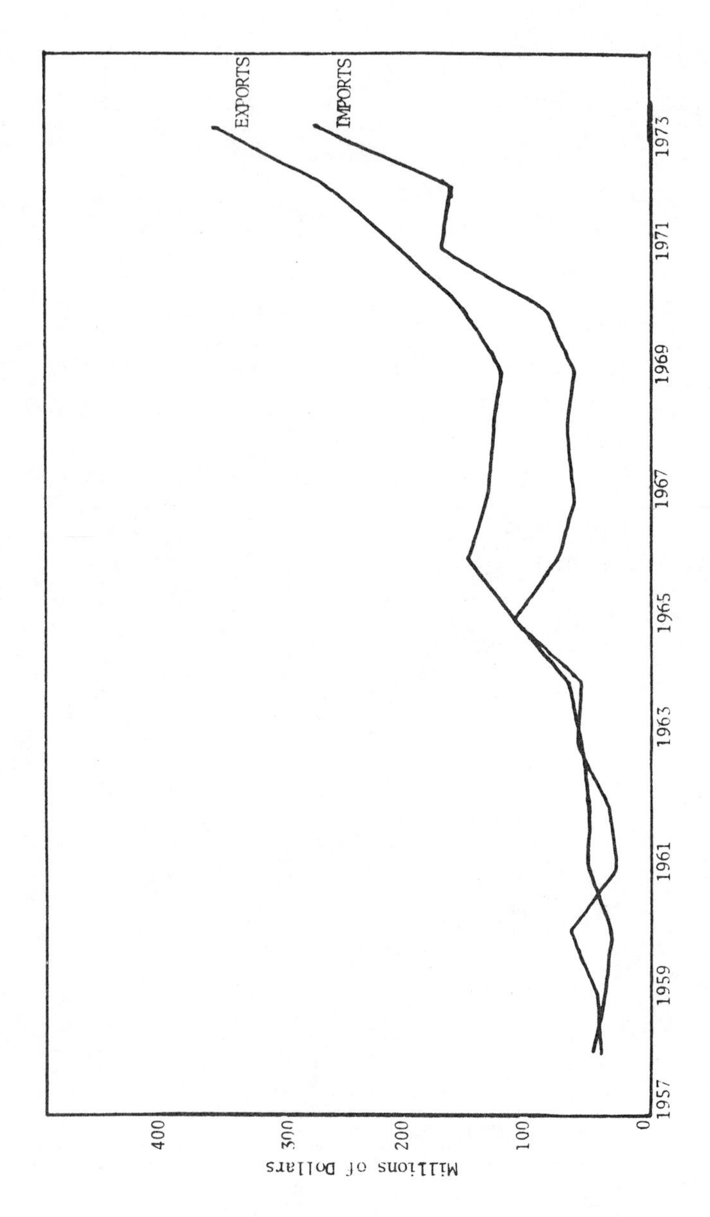

Figure 6. Soviet trade with 42 African countries 1958 through 1974.

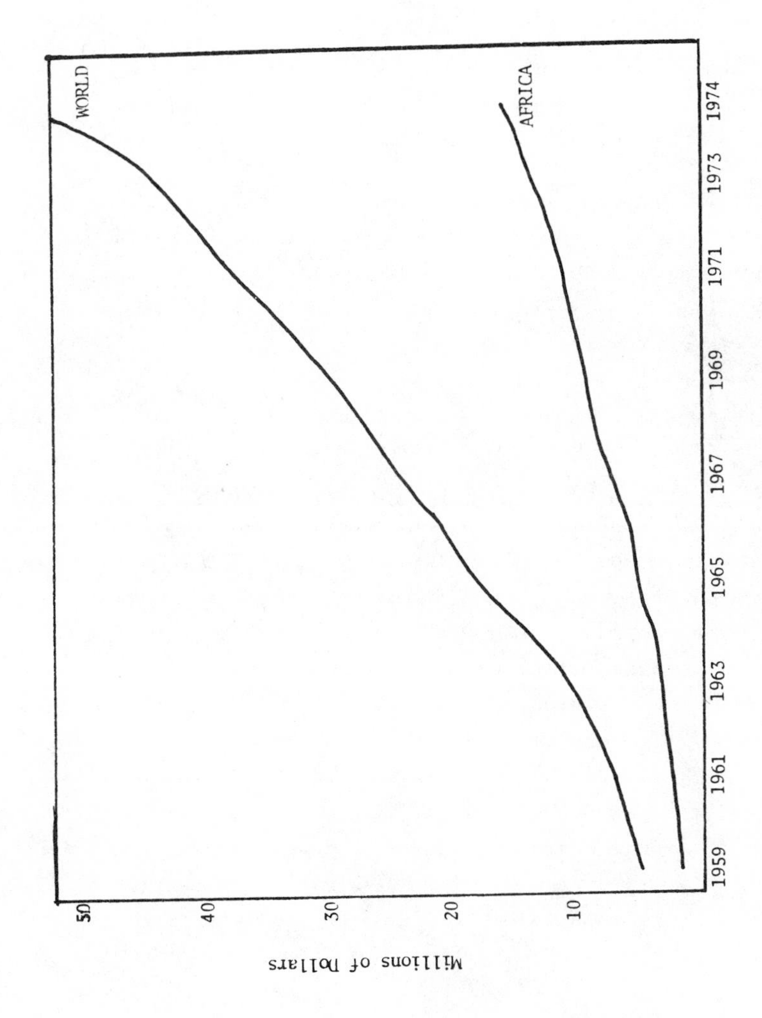

Figure 7. Total Soviet economic aid drawn 1959 through 1974.

Three other variables are included in the analysis to determine whether or not the methodology could indicate that (1) internal characteristics (internal characteristics are variables that describe a particular LDC, e.g., GNP) are considered, or (2) internal characteristics have an influence on the selection of countries as recipients of Soviet foreign policy acts. They are the Gross National Product, Defense Expenditures and Size of the Armed Forces of the targeted country.

The total number of variables for the cluster analysis program is seventeen -- fourteen Soviet foreign policy acts and three internal characteristics.

IV. REPRESENTATION OF MULTIVARIATE DATA WITH COMPUTER DRAWN FACES

The FACES program was developed by Professor Herman Chernoff at Stanford University. The object of the program is to convert multivariate data to a graphic form that can be readily examined. This method of presenting data was chosen under the assumption that the eye functions more like an analog computer than it does a digital computer and is sensitive to changes in facial expression. The various facial characteristics (mouth, nose, eyes, etc.) are modified in form and size to represent statistical vectors.

The programs used are modifications of Chernoff's original program (Chernoff, Personal Communication, 1976). The mofifications include ears and the necessary plotter program changes to operate on the IBM 360 computer with a CALCOMP 765 plotter.

The face drawing program requires twenty numbers to draw a face. Figure 8 and Table 6 describe the features controlled by the parameters $X_1, X_2, \ldots X_{20}$. For example, X_6 controls the nose length and X_{20} the nose width. Table 6 also contains the maximum and minimum range for each of the twenty X's, and the values for the ranges used to plot the faces in Section V. Before proceeding to the modifying instructions, it is necessary to determine the minimum and maximum values for the data set. This can be done quickly for large data sets by using the CONDESCRIPTIVES program in SPSS. After determining these values for each of the variables (Z's), then find which features (X's) will be controlled by which Z's, and find which X's will be fixed. The number of X's to be fixed is 20 - d (d = number of variables). The seventeen variables used to compute the faces in Section V were divided into five groups and targeted as per Table 7. The values of the remaining features must be fixed. To draw a set of faces only the computer input instructions require modification.

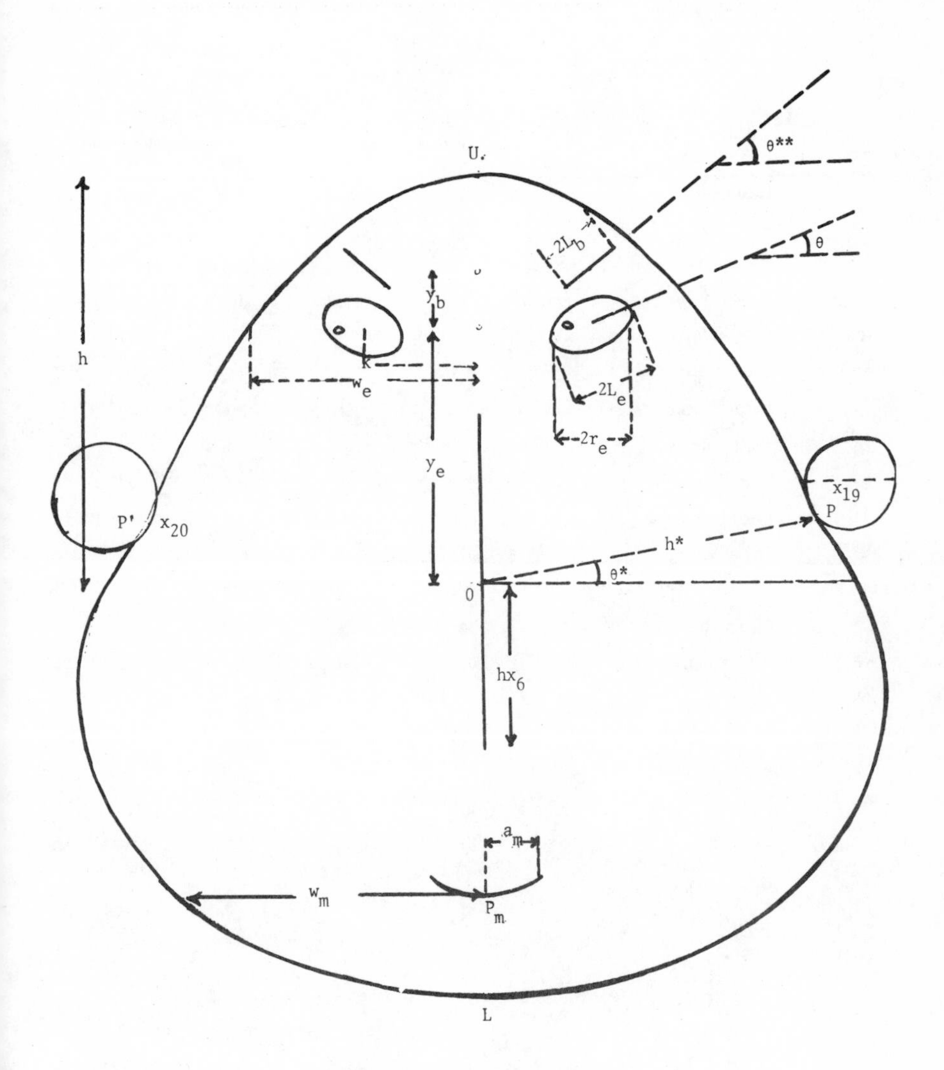

Figure 8. Chernoff face with ears.

Feature Variable	Feature Controlled	Feature Range	Selected Range	Variance Range Min.	Max.
X_1	Face width	0,1	.1,.7	0.0	5.64
X_2	Ear level	0,1	.1,.7	0.0	1,000
X_3	Face height	0,1	.1,.7	5.500	18.04
X_4	Eccentricty upper elipse face	.5,2	.7,1.7	0.0	4.73
X_5	Eccentricity lower elipse face	.5,2	.7,1.7	0.0	2.829
X_6	Nose length	0,1	.2,.9	3.005	16.52
X_7	Mouth center (level)	0,1	.2,.8	0.0	7.0
X_8	Mouth curvature	-5.5	-5,5	0.0	5.32
X_9	Mouth length	0,1	.2,.9	0.0	4.23
X_{10}	Eye level	0,1	Fixed (0.3)	----	----
X_{11}	Eye separation	0,1	.2,.8	0.0	5.84
X_{12}	Eye slant	0,4	.1,.8	0.0	4.873
X_{13}	Eccentricity of eyes (shape)	.4,.8	Fixed (0.6)	----	----
X_{14}	Half length of eye	0,1	.4,.8	0.0	4.71
X_{15}	Pupil position	0,1	.2,.9	0.0	5.51
X_{16}	Eyebrow height	0,1	Fixed (0.5)	----	----
X_{17}	Eyebrow slant	0,1	.2,.6	0.0	3.761
X_{18}	Eyebrow length	0,1	.4,.8	0.0	31.64
X_{19}	Ear diameter	0,1	.1,.5	0.0	9.0
X_{20}	Nose width	0,1	.2,.8	1.00	4.41

Table VI. Parameters for the 20 statistical vectors.

VARIABLE NAMES	FEATURE NUMBER	FEATURE DESCRIPTION
PEOPLE CONTACTS		
Academic Students	X_2	Ear level
LDC Technicians Departing	X_{17}	Eyebrow slant
Soviet Economic Technicians in the LDC	X_{18}	Eyebrow length
MILITARY PERSONNEL (MIL. PERS) CONTACTS		
Soviet Mil Pers in the LDC	X_1	Face width
LDC Mil Pers in the Soviet Union	X_4	Upper elipse of face
Port Visits by Soviet Naval Units	X_5	Lower elipse of face
ECONOMIC CONTACTS		
Economic Agreements	X_{19}	Ear diameter
Imports	X_{11}	Eye separation
Exports	X_{12}	Eye slant
Economic Aid Extended	X_{15}	Pupil position
Yearly Economic Aid Drawings	X_{14}	Half-length of eye
MILITARY AID CONTACTS		
Military Aid Agreements	X_7	Mouth level
Military Aid Extensions	X_8	Mouth curve
Military Grants and Discounts	X_9	Mouth length
LDC CHARACTERISTICS		
Gross National Product	X_3	Face height
Size of Defense Budget	X_6	Nose length
Size of Armed Forces	X_{20}	Nose width
FIXED FEATURES		
Fixed	X_{10}	Eye level
Fixed	X_{13}	Eye shape
Fixed	X_{16}	Eyebrow height

Table VII. Variable association.

V. OTHER APPLICATIONS OF THE FACES PROGRAM

Chernoff published his first article on graphic representation of multivariate data using faces in the Journal of the American Statistical Association in 1971. Following the publication of this article, a number of people working in various fields used this program. Some of these efforts are mentioned below.

Lawrence A. Bruckner of the Los Alamos Scientific Laboratory of the University of California, working under the auspices of USERDA (United States Energy Research and Development Administration), is studying the performance of ten offshore oil groups (Bruckner, 1976).

In 1973 Chernoff applied his methodology to a geological experiment in which he used faces to represent the mineral contents from fifty-three equally spaced samples from a 4500-foot core drilled into a Colorado mountain (Chernoff, 1973).

Probably the most innovative follower of Chernoff's work is Bud Goode of Bud Goode's Sports Computer in Studio City, California. Goode has established his own sports consulting service for major league sports and provides his services to several broadcasting networks who use his faces to predict the results of televized football games. A significant contribution which Goode made to the faces concept was his use of a full character (football player) to represent the data rather than using just the faces.

Goode's latest project is representing trends in U.S. Supreme Court decisions. Future work will include modeling of Congress and of an entire newspaper. He believes that computerized statistical graphical methods will provide useful insight in many areas as long as the messages are simple, direct and can be quickly grasped (Goode, personal communication, 1976).

Another pioneer in computer graphics is Dr. Carol M. Newton of UCLA Health Sciences Computing Facility. Dr. Newton and Jerry Johnson, senior systems programmer for the facility, have developed an interactive graphic (on-line) program in which the analyst works from a video display screen and experiments with his data (Newton and Johnson, interview, 1976).

In addition to the Chernoff faces, Newton and Johnson have developed on-line programs that perform statistical analysis with polygons, bar graphs, arrows and scattergrams. The work has been used in heart research, study of respiratory disease and comparative anthropology.

Johns Hopkins University has begun a number of projects using Chernoff faces. These projects are the result of research done primarily by William H. Huggins, former chairman of the Electrical Engineering Department, in his research on iconic communications. His primary work is in support of a project with the United States Public Health Service Hospital to develop a method of psychiatric screening (Rousuck, 1974).

Dr. David L. Huff of the University of Texas is the most recent person to use the Chernoff faces. He plans to use the methodology to develop urban regional indicators that measure quality of life.

VI. A FOREIGN POLICY APPLICATION OF THE FACES PROGRAM

This section presents the results of the FACE drawing program as applied to the twenty-five country Sub-Saharan African File. Section A takes a look at the long-term trends from 1964 through 1975. The faces represent a four-year average for the country they represent. Section B looks at the changes in the faces in twenty of the countries from 1968 through 1975. Five of the countries which had minimal contact with the Soviet Union were omitted.

A. Soviet Policy, 1964 Through 1975

Figure 9 is a representation of the three families (groupings based on similarity of features) of Soviet foreign policy acts into which the twenty-five countries are grouped. Group one of Figure 9 is constructed using the maximum values for the fourteen foreign policy acts selected for this experiment. The three internal values represent the average for all twenty-five countries. Similarly, group two is constructed with the average values and group three with the minimum values. As with group one, the internal variables were constructed with the mean values.

The validity of using intentional targeting, rather than random targeting, was assessed by a sample test of the data. This assessment was accomplished by giving two sets of twenty-five faces each, one set with the variables intentionally targeted and the other randomly targeted, to two independent subjects and asking each of them to arrange each of the two sets into three families. The results of the two tests were then compared and the findings were that fewer grouping errors occurred in the intentionally targeted sets and the groupings compared with the three faces in Figure 9. Figure 10, which represents Soviet relations with

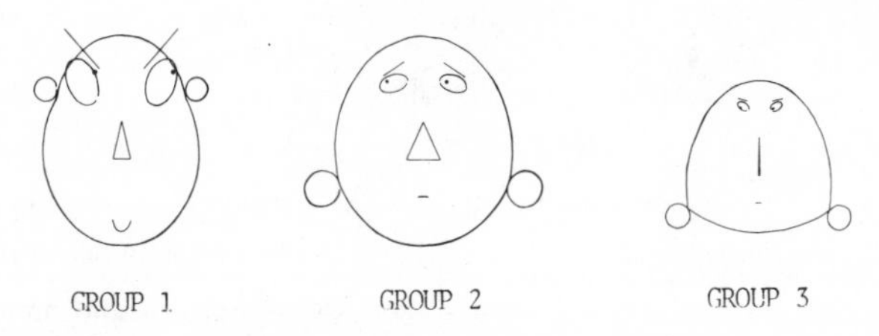

Figure 9. Family groups of Soviet foreign policy acts.

the twenty-five Sub-Saharan African countries from 1964 through 1967, illustrates the data used in this test. It was therefore decided to use the intentional targeting method over random targeting and the faces in this chapter were drawn as previously stated in Section IV.

A comparison of Figures 10, 11 and 12 revealed some interesting facts concerning Soviet foreign policy: 1) an increase in Soviet efforts over the whole area, rather than in selected areas, was indicated by the tendency of the faces to fall into Group Two; 2) Sino-Soviet competition in Africa was supported by two look-alike faces moving from Group One to Group Two in all three figures. A check of the data verified that these two countries (Uganda and Tanzania) had been recipients of aid from the PRC and the USSR.

The expressions of the three countries (faces with worried looks) in Group One of Figure 10 (People's Republic of the Congo, Uganda, and Tanzania), are influenced by the economic contacts and people contacts shown in Table 7 (Section IV). The ear diameter and pupil position indicate a willingness for economic involvement on the part of the Soviets, but the eye and eyebrow indicate that little follow-through of initial effort is apparent. The pupil position and ear height of Kenya in Group Three of Figures 10 - 12 provide an early economic indicator. The country showing the most upward mobility between Figures 10 and 11 is Sudan. But in Figure 12 the change in face width, indicates that the number of Soviet military personnel decreased in the final time period.

In summary, Figures 10, 11, and 12 appear to indicate the existence of a stable policy overall in the region. Some evidence of the Sino-Soviet competition is apparent with an loss for the Soviets. The three countries in Group One in Figure 12 are in Group One in Figure 10 and 11, also.

B. Selected Yearly Faces from 1968 through 1975

Figures 13 through 19 represent the faces of Soviet foreign policy in twenty countries on a yearly basis from 1968 through 1975. Each country is represented by two rows of faces, reading from left to right for the years 1968 through 1975. There are three countries per figure except for Figure 14 which represents only two countries, Nigeria and the People's Republic of the Congo. Figure 13 presents the faces with the highest values and Figure 19 presents the faces with the lowest values for the set.

Figure 13 shows a Soviet pattern present in Guinea, Somalia and Mali Republic. In Each of the three sets of faces, the length of the eyebrow (number of Soviet economic technicians in the country) is followed by a change in the angle of the eyebrow (number of LDC technicians departing for training in the USSR) then by a change in pupil position (economic aid extended). Another indicator is in the relationship between the angle of the eyebrows and the eccentricity of the forehead (LDC technicians and military personnel departing for the Soviet Union). Until the angle of the left eyebrow reaches a horizontal plane in clockwise rotation, it appears to be a tangent of the eccentricity of the upper

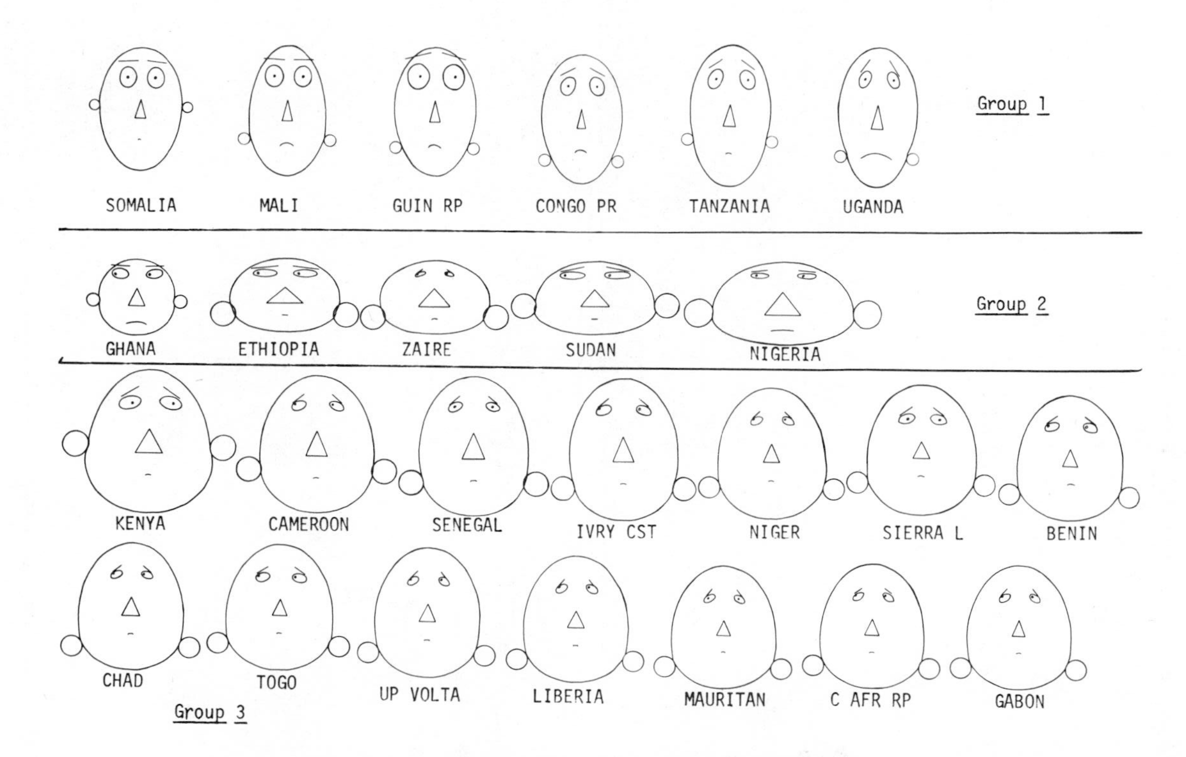

Figure 10. Soviet foreign policy faces 1964 through 1967.

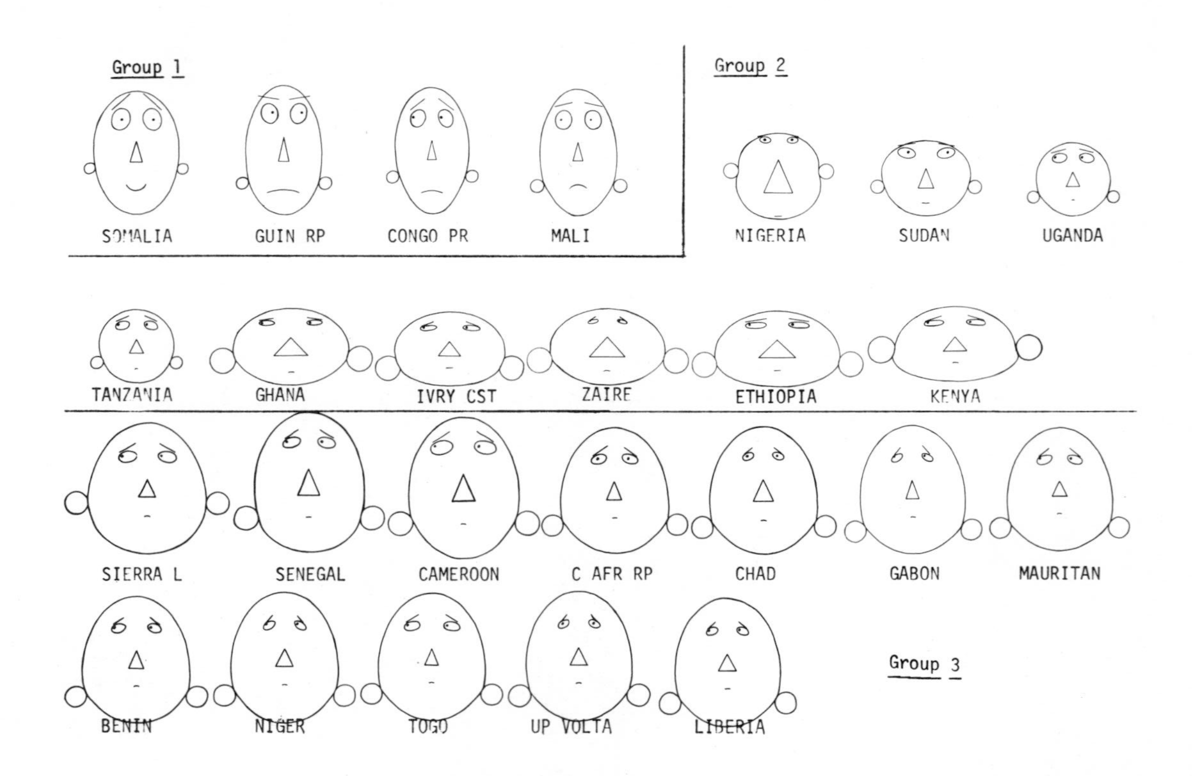

Figure 11. Soviet foreign policy faces 1968 through 1971.

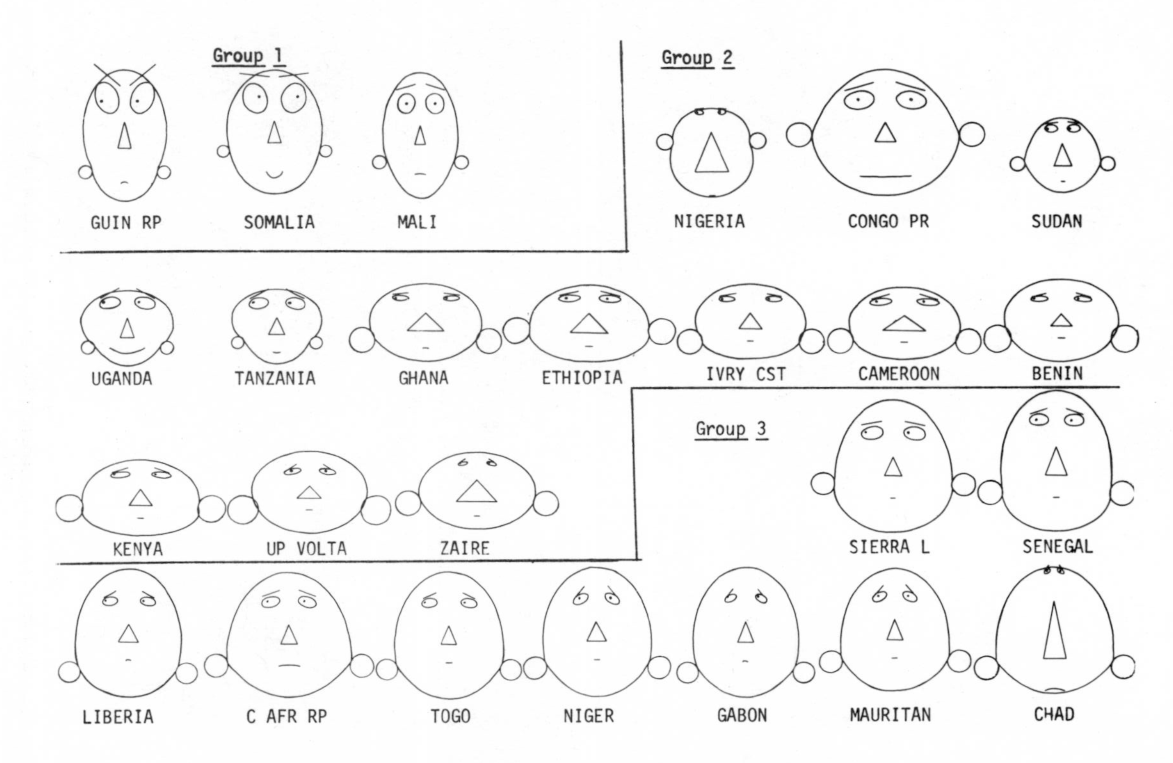

Figure 12. Soviet foreign policy faces 1972 through 1975.

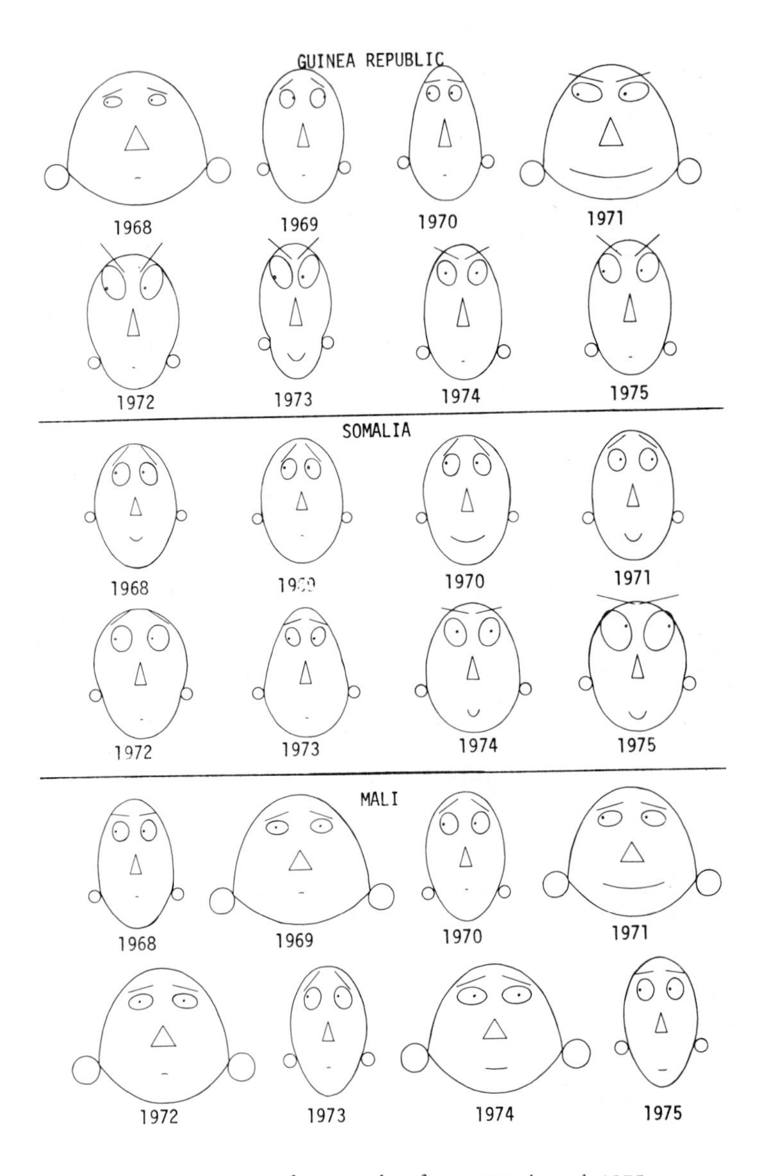

Figure 13. Soviet foreign policy faces 1972 through 1975.

ellipse of the face. A long smiling mouth (curvature is controlled by value of military aid extension--the more the happier, and mouth length is controlled by the size of grant or discount) is a relationship that only occurs in four countries, the three in Figure 13 and the People's Republic of the Congo in 1975 (Figure 14). It is also usually preceded by a decrease in ear diameter (number of economic agreements) and a lowering in the height of the mouth (increase in number of military aid agreements).

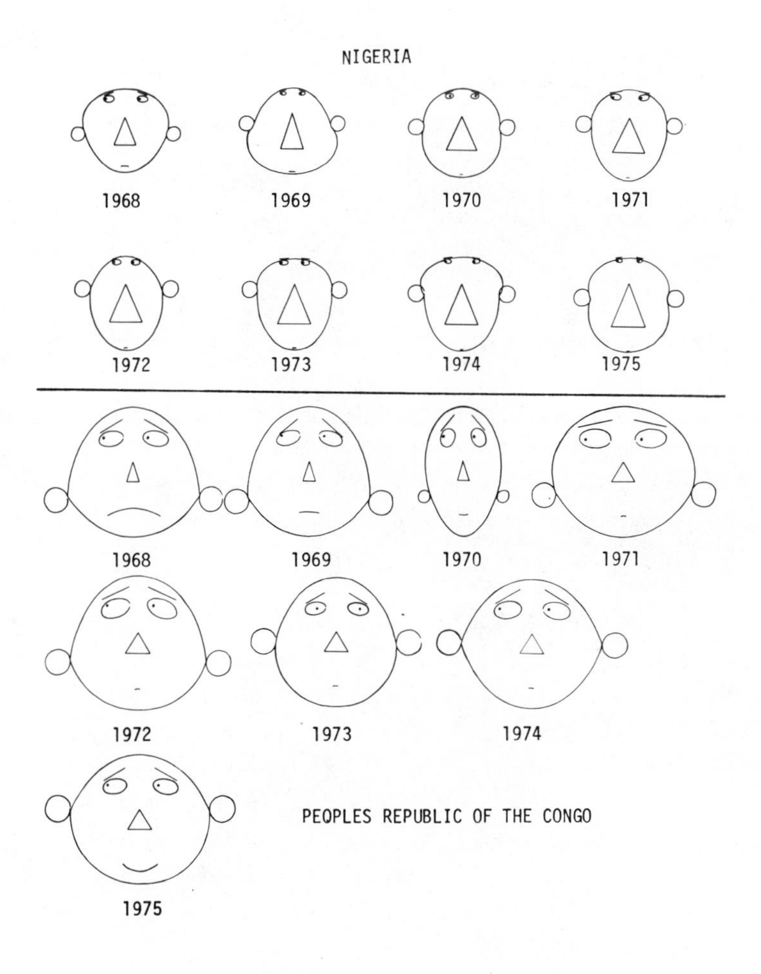

Figure 14. Changing faces 1968 through 1975.

Figure 14 presents two of the more interesting groups of faces, Nigeria and the People's Republic of the Congo. Nigeria has all the early indicators, i.e., large number of students in the Soviet Union (ear height), economic aid agreements (ear diameter), several military aid agreements (as the number of aid agreements increase, the mouth level drops) and Soviet military personnel present (face width). Also noticeable is the eccentricity of the chin (increasing number of port visits) for 1969, 1970 and 1975. In 1968, 1974 and 1975 the upper ellipse of the face indicates substantial numbers of Nigerian military personnel departing for training in the Soviet Union. Still these faces present a strange appearance. This uniqueness is explained by the eyebrows, eyes and the nose. Economic aid and trade is low (eyes and eyebrows), GNP is high (face height, as values increase, height decreases) and it has a large defense budget and large armed force (nose height and width respectively). The year 1971 indicates, by pupil movement, that substantial economic aid grant was offered, but there was no indication of acceptance.

The People's Republic of the Congo shows an overall increased Soviet contact with a slight setback in economic relations in 1973. The smiling face of 1975 would be the one of most concern to an analyst as it indicates large grants and discounts as well as large military aid extensions and several port visits.

The first set of faces in Figure 15 indicates that Soviet presence in Sudan was well-established in 1968. The smile in 1968 shows a large military aid extension (mouth curvature), but no discounts (mouth length). Other 1968 indicators are large economic aid drawing (eye slant) and large number of Soviet economic technicians (eyebrow length). A large economic extension (pupils), increase in students (ear height) which reaches its height in 1970, and port visits by Soviet naval units (chin) are shown in 1969. In 1971 a turn for the worse for the Soviets begins to appear. The eyebrows start a counter-clockwise rotation followed in 1972 by a similar pattern with eye slants. In 1974 a sharp increase in the number of Soviet technicians occurs in an apparent effort to save its program in Sudan. This was followed in 1975 by a large withdrawal of Soviet military personnel.

The second set of faces in Figure 15 are those illustrating Soviet penetration in Uganda. A period of steady economic growth begins in 1970 for the Soviets above the mean value for the data set. The eyebrows and eyes begin their rotations in 1970 and continue even through 1971 and 1972 when there is a decrease in the number of Soviet military personnel (face width). The period 1973 through 1975 is a time of steady gains for the Soviets in Uganda. The four features with the most significant change are: ear diameter (economic aid agreements); half length of the eye (yearly economic aid drawing); eyebrow length (increase in the number of Soviet economic technicians); and face width (increase in number of Soviet military personnel). The large smile in 1974 indicates Uganda received the most difficult of the foreign policy acts–a large military extension combined with a substantial grant or discount.

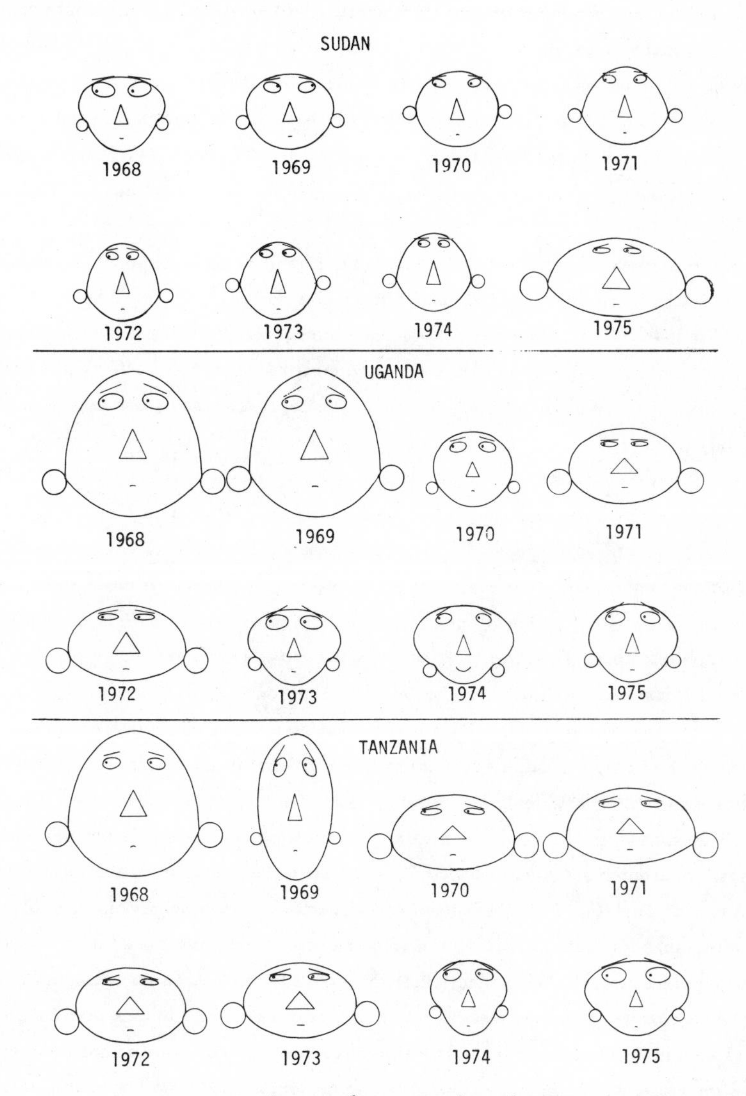

Figure 15. Changing faces 1968 through 1975.

The final set of faces in Figure 15 illustrates Soviet penetration in Tanzania. As previously discussed in Section A of this section, Tanzania and Uganda present a similar pattern of Soviet activities. Figure 15 presents the similarity more clearly. In 1969 the upper ellipse of the face shows a large number of military personnel departing for the Soviet Union which drops off again in 1970. Also indicated in 1970 is a decrease in the number of port visits. An increase in Soviet economic technicians is indicated in 1969. This increase establishes a trend for this variable that continues through 1975. A decrease in the number of Soviet military personnel is shown from 1970-1974. The Soviet face for Tanzania in 1974 is very similar to the Soviet face for Uganda in 1973. These two latter faces indicate a possible trend which can be stated roughly as follows: when relations are restored after a setback, they rapidly return to and often exceed the level of prior activity.

The next six sets of faces shown in Figures 16 and 17 all have moderate relations with the Soviet Union. Some effort of penetration is apparent in Ghana, Figure 16, through changes in ear level, eye shape, eyebrow length and the structure of the mouth. The face for 1975, however, reveals that little success has so far resulted from the efforts. The second set of faces in Figure 16, Ethiopia, shows a more favorable face each year towards the Soviets. The only noticeable change in the Ivory Coast, Figure 16, is in imports, exports and the number of naval port visits.

Cameroon, shown in Figure 17, demonstrates some activity in the eye and eyebrow. A slow increase in the number of economic technicians is paralleled by a similar increase in trade and economic aid drawings. A shift to the left in pupil position in 1974 signals an economic aid extension. Future Cameroon faces should be watched for changes in the military related variables. The bottom set of faces in Figure 17 is Kenya. The trend for Kenya is an overall decrease in Soviet activity with little military activity, except for port visits.

Figure 18 represents Upper Volta, Zaire and Senegal. Changes in Upper Volta are not apparent until 1973. In addition to an increase in GNP to above the mean level, the ears show a slight rise (increase in the number of students in the Soviet Union) and the eyes being to open more to the Soviet view (pupil movement to the left which reflects an increase in economic aid extended and change in the eye shape caused by an increase in imports and exports). The only activity indicated in Zaire is in the number of students attending school in the Soviet Union. Senegal, however, shows a definite potential for increased Soviet involvement. The ellipse of the chin indicates numerous port visits each year. The ears are slowly rising, reflecting an increase in the number of students in the Soviet Union, and the variables in and around the eyes demonstrate an increasing trend in economic activity.

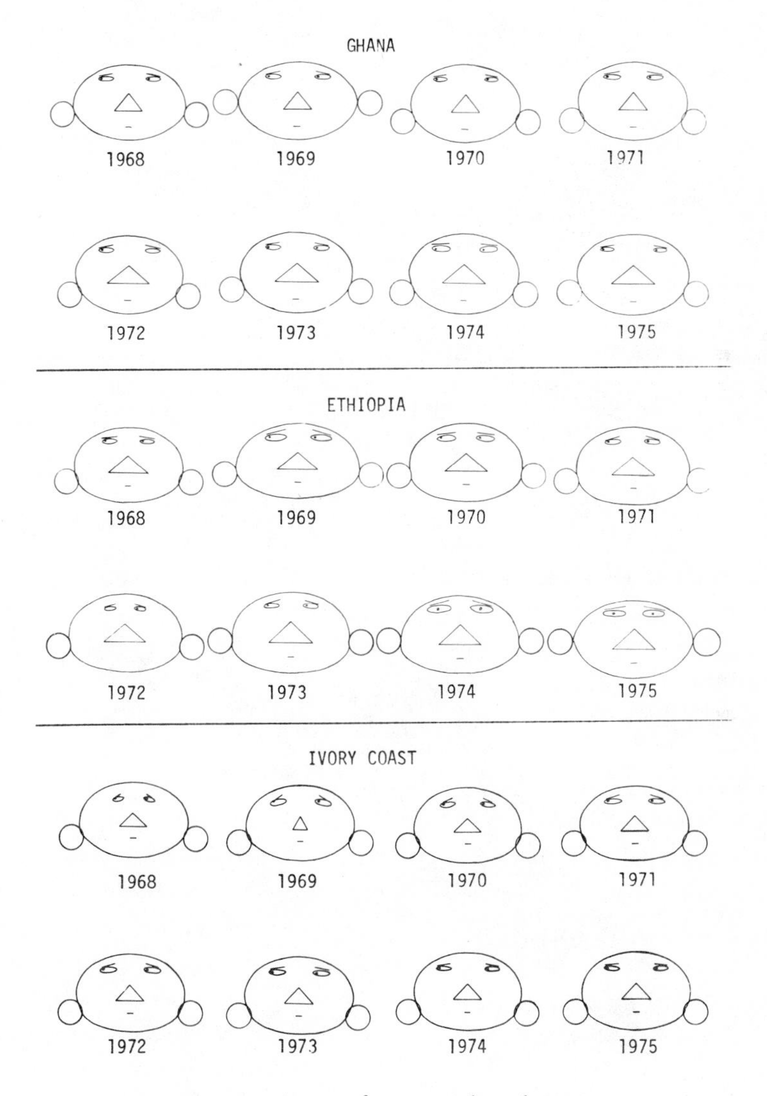

Figure 16. Changing faces 1968 through 1975.

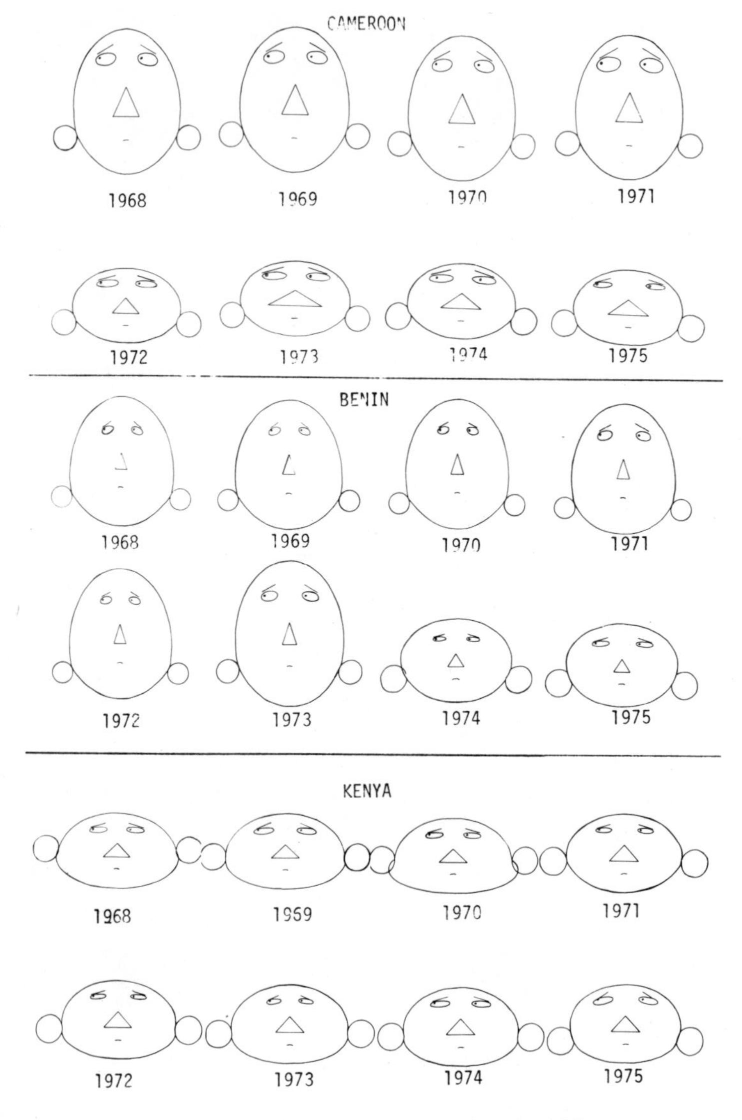

Figure 17. Cnanging faces 1968 through 1975.

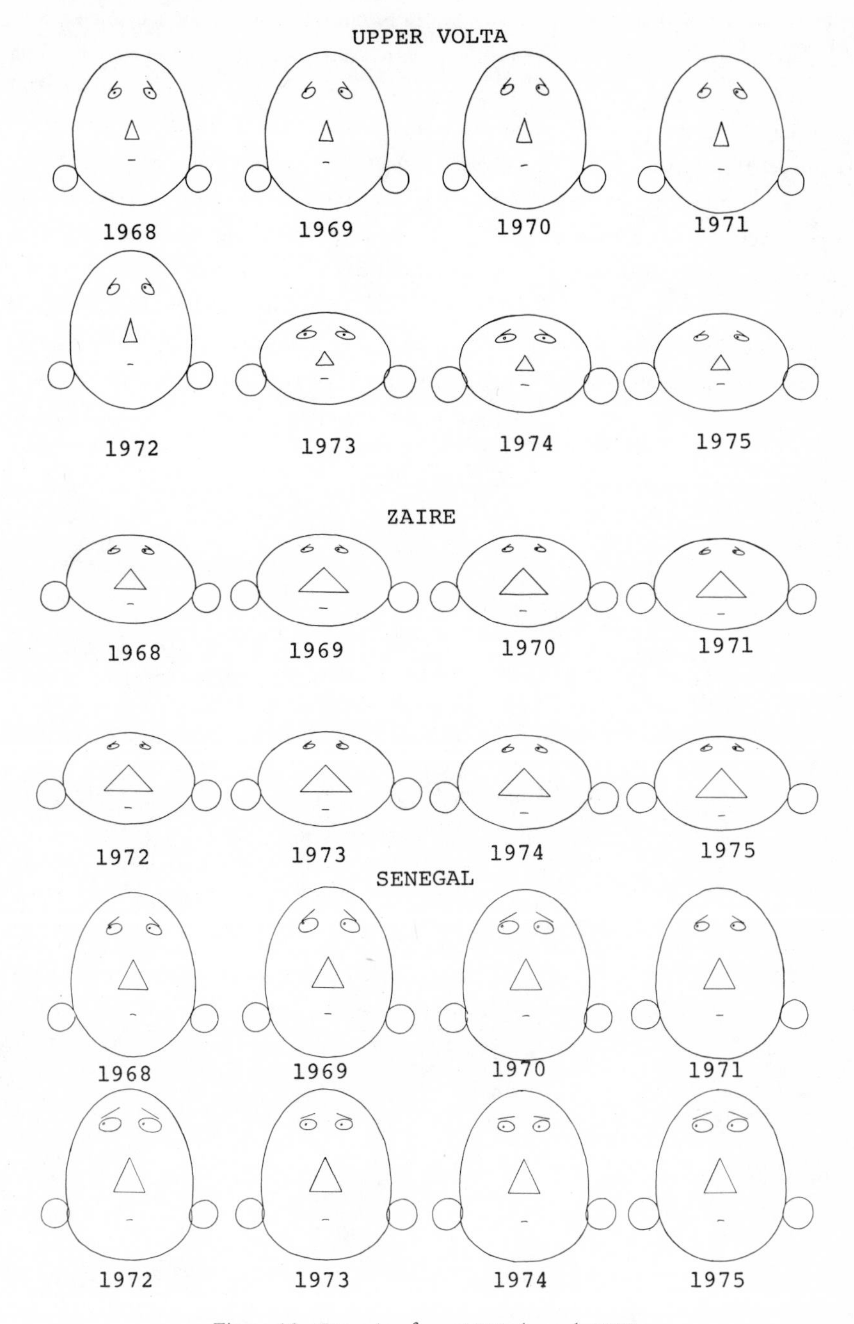

Figure 18. Cnanging faces 1968 through 1975.

Figure 19 represents Soviet foreign policy in Liberia, Gabon and Chad. The ear level for Liberia rises throughout the range of faces (increases in the number of students from Liberia in the Soviet Union). The beginning of trade activity opens the eyes in 1974 (increase in imports and exports change eye separation and eye slant). Gabon represents a country with minimum contact for all eight years. Chad indicates some economic offering (pupil shift to the left and then back to the center) starting in 1973, economic technicians arrive (length of the eyebrows increase). The most noticeable change occurs in 1975, when the mouth length increases (large military grant or discount). It is interesting to note that the grant occurs during the same year when the Soviets are experiencing difficulty in neighboring Sudan. Experimenting with a larger data base could reveal possible Soviet military geographical interest and associated patterns of penetration.

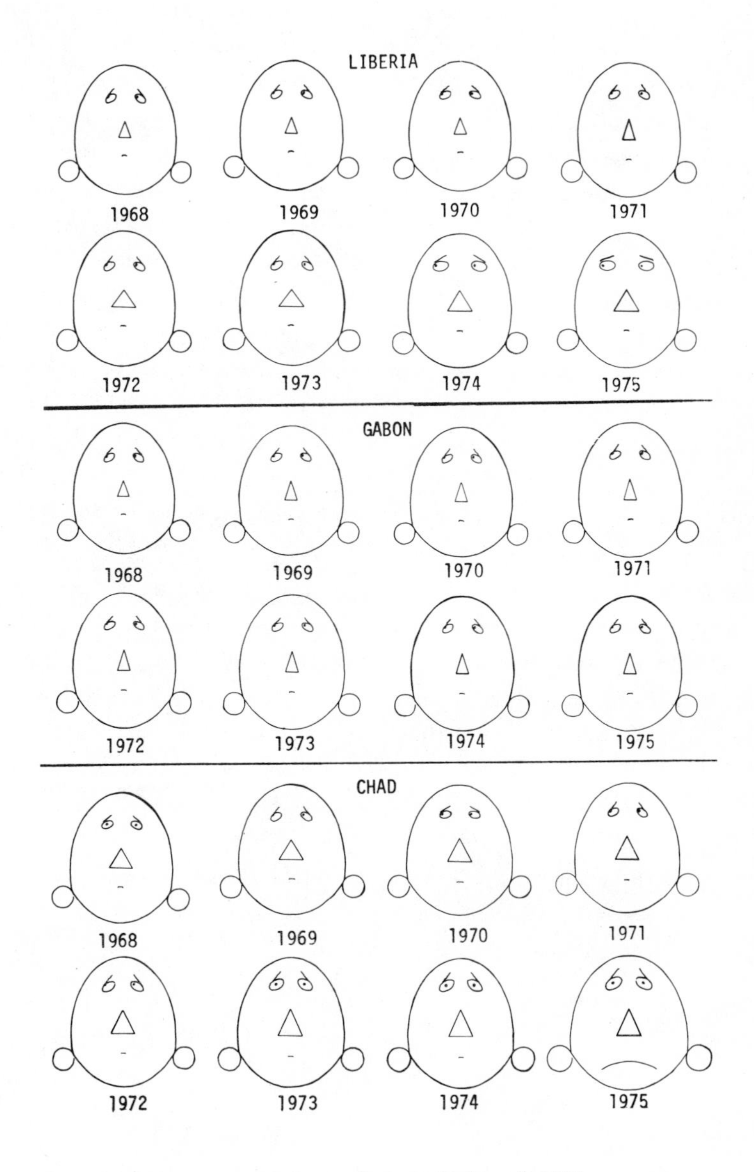

Figure 19. Changing faces 1968 through 1975.

VII. CONCLUSION

The utility of Chernoff's faces as a diagnostic or evaluation tool demonstrates a promising approach for a first look at multivariate data. It has the advantage that in analysis of complex relations, indications can be made of patterns or clusters that are not always visible from simple correlations based on two-dimensional linear theories.

As international relations become more complex and the probability of Third World nations' influence on the course of international relations continues, a growing need emerges for real time systems to monitor, display and communicate the complex phenomena associated with international relationships. At present the majority of analyses of international relations is ex post facto. With increasing means of transportation and communication, the role of the analyst must change from an emphasis on evaluating what has already happened to an increasing responsibility to forecast, with degrees of confidence, the probable outcome of events, and to recommend alternatives in foreign policy, deployment of military units and procurement of weapons systems. These forecasts will enable decision makers to make the best selection of the alternatives available to yield desired results.

Other types of computer graphics displays can be developed to aid analysts in modeling Soviet foreign policy. As demonstrated in the FACES program, the variables can be represented by a vector with length and motion. By using three-dimensional graphic equipment, the variables could be represented by a series of ellipsoids. Where $X^2/a^2 + Y^2/b^2 + Z^2/c^2$ could represent the properties of each variable and it could be rotated; the velocity of rotation would represent the rate of change for the variable. Sets of these ellipsoids could be displayed simultaneously and through time-series analysis, the analyst could visually examine the phenomena. The primary reason for this approach is that the variables are multi-dimensional and the real relationships are therefore compound multi-dimensional phenomena. If this technique still indicated patterns between or within the ellipsoids, consideration could be given to applying the basic laws of nature for frameworks for modeling.

Computer-aided graphical display of international relations phenomena has the potential for moving international studies and analysis from an art to a science, which allow for a rapid simplistic presentation of complex relations.

APPENDIX A

1. Abbreviations for the 195 Countries in the WARP File

Country Abbreviation	Country Name
AFGHNSTN	AFGHANISTAN
AFRCA NS	AFRICA NS
AFRS-ISS	AFARS-ISSAS
ALBANIA	ALBANIA
ALGERIA	ALGERIA
ANGOLA	ANGOLA
ARGENTIN	ARGENTINA
AUSTRALI	AUSTRALIA
AUSTRIA	AUSTRIA
BAHAMAS	BAHAMAS
BAHRAIN	BAHRAIN
BARBADOS	BARBADOS
BELGIUM	BELGIUM
BELIZE	BELIZE
BENIN	BENIN
BERMUDA	BERMUDA
BHUTAN	BHUTAN
BNGLDSH	BANGLADESH
BOLIVIA	BOLIVIA
BOTSWANA	BOTSWANA
BR SOLMN	BRITISH SOLOMON
BR WI NS	BRITISH WEST INDIES NS
BRAZIL	BRAZIL
BRUNEI	BRUNEI
BULGARIA	BULGARIA
BURMA	BURMA
BURUNDI	BURUNDI
C AFR RP	CENTRAL AFRICAN REPUBLIC
C VRD IS	CAPE VERDE ISLANDS
CAMBODIA	CAMBODIA
CAMEROON	CAMEROON
CANADA	CANADA
CHAD	CHAD
CHILE	CHILE
CHINA ML	CHINA MAINLAND
CHINA RP	CHINA, REPUBLIC OF
CMORO IS	COMORO ISLANDS
COLUMBIA	COLUMBIA
CONGO PR	CONGO, PEOPLE'S REPUBLIC
COSTA RI	COSTA RICA
CUBA	CUBA
CYPRUS	CYPRUS
CZCHSLVK	CZECHOSLOVAKIA
DENMARK	DENMARK
DMNCN RP	DOMINICAN REPUBLIC
E GRMNY	EASTERN GERMANY
EA PC NS	EAST ASIA AND PACIFIC NS
ECUADOR	ECUADOR
EGYPT	EGYPT
EL SALVA	EL SALVADOR

Country Abbreviation	Country Name
EQU GUIN	EQUATORIAL GUINEA
ETHIOPIA	ETHIOPIA
EUROP NS	EUROPE NS
FAROE IS	FAROE ISLANDS
FIJI	FIJI
FINLAND	FINLAND
FLKN IS	FALKLAND ISLANDS
FR WI NS	FRENCH WEST INDIES NS
FRANCE	FRANCE
GABON	GABON
GAMBIA	GAMBIA, THE
GERMANY	GERMANY
GHANA	GHANA
GIBRLTAR	GIBRALTAR
GLB-E IS	GILBERT-ELLICE ISLANDS
GREECE	GREECE
GREENLAND	GREENLAND
GUADLUPE	GUADELOUPE
GUAM	GUAM
GUATEMAL	GUATEMALA
GUIAN FR	GUIANA, FRENCH
GUIN RP	GUINEA REPUBLIC
GUIN-BIS	GUINEA-BISSAU
GUYANA	GUYANA
HAITI	HAITI
HONDURAS	HONDURAS
HONGKONG	HONG KONG
HUNGARY	HUNGARY
ICELAND	ICELAND
INDIA	INDIA
INDONESI	INDONESIA
IRAN	IRAN
IRELAND	IRELAND
ISRAEL	ISRAEL
ITALY	ITALY
IVRY CST	IVORY COAST
JAMAICA	JAMAICA
JAPAN	JAPAN
JORDAN	JORDAN
KENYA	KENYA
KOREA	KOREA
KUWAIT	KUWAIT
LAOS	LAOS
LEBANON	LEBANON
LESOTHO	LESOTHO
LEWRD IS	LEEWARD ISLANDS
LIBERIA	LIBERIA
LIBYA	LIBYA
LUXEMBRG	LUXEMBOURG
MACAO	MACAO
MALAWI	MALAWI
MALAYSIA	MALAYSIA
MALDIVES	MALDIVES

Country Abbreviation	Country Name
MALI	MALI REPUBLIC
MALTA	MALTA
MARTINIQ	MARTINIQUE
MAURITAN	MAURITANIA
MAURITIU	MAURITIUS
MEXICO	MEXICO
MLGSY RP	MALAGASY REPUBLIC
MNGLN RP	MONGOLIAN REPUBLIC
MOROCCO	MOROCCO
MZMBIQUE	MOZAMBIQUE
N KOREA	NORTH KOREA
N VIETNM	NORTH VIET-NAM
NAMIBIA	NAMIBIA
NAURU	NAURU
NE SA NS	NEAR EAST AND SOUTH ASIA NS
NEPAL	NEPAL
NEW CLDN	NEW CALEDONIA
NEW HBRD	NEW HEBRIDES
NEW ZLND	NEW ZEALAND
NICARAGU	NICARAGUA
NIGER	NIGER
NIGERIA	NIGERIA
NORFOLK	NORFOLK
NORWAY	NORWAY
NTHR ANT	NETHERLANDS ANTILLES
NTHRLNDS	NETHERLANDS
OMAN	OMAN
OT AFRCA	OTHER AFRICA
OT EA PC	OTHER EAST ASIA AND PACIFIC
OT EUROPE	OTHER EUROPE
OT NE SA	OTHER NEAR EAST AND SOUTH ASIA
OT W HMP	OTHER WESTERN HEMISPHERE
PAKISTAN	PAKISTAN
PANAMA	PANAMA
PAPUA NG	PAPUA NEW GUINEA
PARAGUAY	PARAGUAY
PERU	PERU
PHLPPNS	PHILIPPINES
PN CNL Z	PANAMA CANAL ZONE
POLAND	POLAND
PORTUGAL	PORTUGAL
QATAR	QATAR
REUNION	REUNION
RHODESIA	RHODESIA
RWANDA	RWANDA
S AFRICA	SOUTH AFRICA
S VIETNM	SOUTH VIET-NAM
SAMOA AM	SAMOA, AMERICAN
SAO T PR	SAO TOME & PRINCIPE
SAUDI AR	SAUDI ARABIA
SENEGAL	SENEGAL
SEYCHLLS	SEYCHELLES
SIERRA L	SIERRA LEONE

Country Abbreviation	Country Name
SINGAPOR	SINGAPORE
SOMALIA	SOMALIA
SPAIN	SPAIN
SRI LANK	SRI LANKA
ST HELNA	SAINT HELENA
ST PR-MQ	SAINT PIERRE-MIQUELON
SUDAN	SUDAN
SURINAM	SURINAM
SWAZILND	SWAZILAND
SWEDEN	SWEDEN
SWTZRLND	SWITZERLAND
SYRIA	SYRIA
TANZANIA	TANZANIA
THAILAND	THAILAND
TIMOR	TIMOR
TOGO	TOGO
TONGA	TONGA
TRND TBG	TRINIDAD AND TOBAGO
TUNISIA	TUNISIA
TURKEY	TURKEY
UGANDA	UGANDA
UN AR EM	UNITED ARAB EMIRATES
UN KNGDM	UNITED KINGDOM
UP VOLTA	UPPER VOLTA
URUGUAY	URUGUAY
US VG IS	VIRGIN ISLANDS, US
USA	UNITED STATES
USSR	USSR
VENEZUEL	VENEZUELA
W HMP NS	WESTERN HEMISPHERE NS
W SAMOA	WESTERN SAMOA
WINDW IS	WINDWARD ISLANDS
YEMEN AR	YEMEN ARAB REPUBLIC
YEMEN PR	YEMEN, PEOPLE'S REPUBLIC
YUGOSLAV	YUGOSLAVIA
ZAIRE	ZAIRE
ZAMBIA	ZAMBIA

2. Schematic Diagram of the WARP File

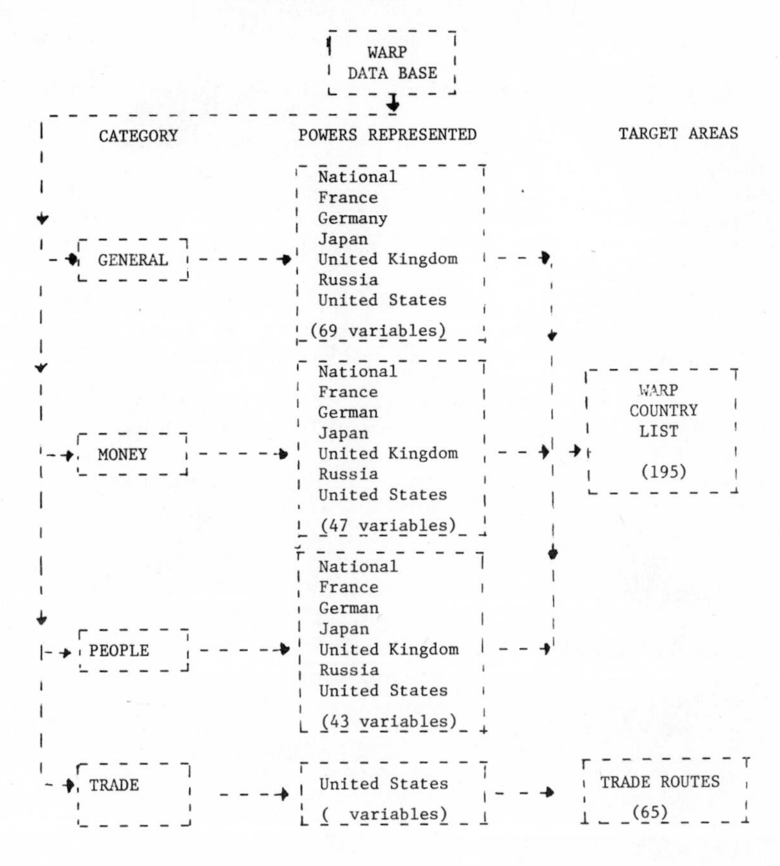

REFERENCES

Bruckner, L. (1976). "The Looks of Some Companies Involved in Offshore Oil and Gas Leases." Paper presented at the annual meeting of the American Statistical Association, Boston.

Center for Naval Analyses (1976). "Orchestration Codebook." Arlington, Virginia.

Chernoff, H. (1971). "The Use of Faces to Represent Points in n-Dimensional Space Graphically." Stanford University Department of Statistics, Technical Report No. 71.

Chernoff, H. (1973). JASA. 68:361.

Christensen, J. and Pieper, J. (1976). "World Analytic Research Project (WARP): Data Base Documentation and User's Guide." Air Force Data Services Center, Washington.

Gurr, T. (1972). "Politimetrics: An Introduction to Quantitative Macro-Politics." Prentice-Hall, Inc., Englewood Cliffs.

Lindsey, G. (1976). Lecture notes from 0A4207 Scaling Techniques. Naval Postgraduate School, Monterey, California.

Nie, N. et al. (1975). "SPSS: Statistical Package for the Social Sciences," Second Edition. McGraw-Hill Book Company, New York.

Rousuck, J. (1974). "Computer Faces that 'Talk'." The Sun Magazine, August, p. 12.

Sherwin, R. (1973). "WEIS Project Final Report." UCLA Press, School of International Relations, Los Angeles.

GRAPHICS: FROM ALPHA TO OMEGA IN DATA ANALYSIS

Carol M. Newton

Department of Biomathematics
University of California
Los Angeles, California

I. INTRODUCTION; FROM ALPHA TO OMEGA AND BACK

Our curiosity is struck by some pattern in the world about us. We poke at it to see what happens. We turn it around to examine it from various perspectives. We form some thoughts about what it might mean, about artifactuality vs. actuality, about relationships to other things we know. We then phrase hypotheses and begin the exacting task of testing them. Many fail. Some survive to be further tested, and some of these survive to be knit into the structure of our scientific knowledge, from which future investigations can reliably take departure. But often it doesn't end here. Increasingly, fortunately, decisions in our complex practical undertakings seek foundations in scientific knowledge. The decision and policy makers enter the scene as new learners of established relationships. Perhaps what helped the scientists to first perceive and study these relationships will help the new people to comprehend those relationships that have survived to become knowledge, --- and to better visualize how they might apply to their practical problems.

ISBN 0-12-734750-X

It is the contention of this paper that at the alpha and omega of our knowledge process, --- from first intuition to the final handoff of knowledge to those who seek to apply it on behalf of mankind, effective graphical aids are of increasingly vital importance as the complexity and difficulty of our scientific and applied undertakings progress. But such aids can also guide us in the more formal hypothesis-testing aspects of research as well, and would be indispensable components of an ideal interactive system that comprehends data analysis as a continuous, recycling activity.

II. THE ALPHA

We can disagree about what we may want to call it, exploratory data analysis or something else, but we can welcome the increased attention that is being paid to aids to our earliest examination of scientific data. Statistics traditionally has served very well the domain of scientific activity that centers on hypothesis testing, and well it might. Statistical research evolved when there was a compelling need to place knowledge on a firm footing, and the statistician's role evolved as a developer of the relevant methodologies of experimental design and formal analyses, and as collaborator in other research activities requiring such methodological supports. Although statistical attention certainly was given to it, the domain of scientific discovery --- the inductive leap, the first "a-*ha*!" --- was pretty much left to the scientist himself. And it should be conceded that his may remain the most essential contribution in that domain. The chemistry is as yet poorly understood of interactions between a scientist's extensive scholarly understanding of a field and of its possible underlying mechanisms, his perception of patterns that still would elude formal recognition techniques, and his innate intuitive inventiveness in proposing new relationships that cannot be derived by deductions from the existing body of scientific knowledge (and hence are essential to its growth). What we can do right now is to seek to find aids to this process, and to place them effectively in the hands of the people who are best able to pursue it. The latter might include statisticians who are well versed in the field being studied, but what I have more in mind are the investigating scientists themselves, by themselves or in the company of a statistician. Knowing what might be going on underneath the data is important to ferreting out scientifically useful patterns, --- the difference between astrology and astronomy.

Until scientists are more adequately trained in the mathematics and statistics that is relevant to their work, the advantage of being able to convey relationships to them by graphical means is self evident, --- and such means also are attractive to people who are mathematically trained. The human eye is a marvelous device for discerning patterns in complex settings, as many a computer scientist working in the area of pattern recognition has learned. Accepting that we shall seek to provide good graphical aids to scientists in their early examination of data, what are some of the issues to consider? The following come immediately to mind, and very likely there are more:

(1) the form of graphical representation

(2) the method of presentation

(3) general vs. specialized graphics supports

(4) costs, feasibility

I shall defer discussion of the last point until a later section of the paper.

A. The Form of Graphical Representation

Innovative devices for summarizing data are an object of serious investigation today. Computer graphics makes their construction economically and temporally feasible for routine use. Chernoff's faces (1), multivariate polygons (2), etc. come readily to mind. But one also can concentrate on the more effective use of traditional graphics devices, e.g. associating a set of scattergrams in order to escape the three-dimensional barrier in examining relationships among individual cases (3).

An issue that one hears argued is that of how "unbiased" the form of representation should be. Some people will accept a graphical form of representation only if it summarizes information without any possible distortion. Such people argue against the "faces" because, for instance, a subjective bias might be introduced on the basis of which variables are assigned to which features (e.g. the face smiles for a given subset of cases, inclining the viewer to be better disposed toward them). These concerns have some merit when we reach omega, but far less during the process of discovery where in fact some distortions might attract attention to relationships that should be studied further.

The entire point of the discovery process is to bring to attention relationships of possible interest; the subsequent process of experimentation and hypothesis testing is relied upon to wash out spurious relationships.

B. The Method of Presentation

The major division here is along the lines of interactive and passive graphics. In the latter, one examines graphical output that has been prepared by the computer at another time. One can seek patterns in a large set of printed scattergrams and their associated statistical summaries. One can cut out the faces that have been produced by a plotter, and then proceed to assign them to groups having similar appearances. After this examination of output has been completed, one may ask for another set of graphical output to be produced. It is quite possible to effectively explore data this way; many people do. If one's attention is on computing costs, this approach might be cheaper provided that the investigator doesn't order far more than he needs at a given step because he doesn't want the bother and delay of submitting a sequence of many jobs to the computer. If one's attention is on total costs or on accelerating the discovery process, where the investigator's time and salary are matters of importance, interactive graphics may have the cost advantage.

Interactive graphics provides the investigator a set of graphical and analytical tools, retrieval capabilities, etc. and places him in direct dialogue with his data base. He can move freely from one display to another, examining raw data for outliers, looking for the persistence of clusters in an associated set of scattergrams, rotating a scattergram to expose a third dimension, comparing groups or individuals by "faces," polygons, or other summary devices. Here the problem of bias can be minimized by providing the investigator complete freedom to rearrange the assignment of variables to features, and encouraging him to exercise it. He will in fact find that certain features are more influential than others when we classify faces, and that the classification of cases into various groups might be altered if he changes the assignment of variables to features. Seeing how invariant a perceived pattern is to various representations and in different dimensions is a first step toward eliminating attention to spurious relationships.

Couldn't all of this be done in the passive mode? Much of it can, though far less effectively. When a scientist is engrossed in following up on the hunches that come to mind as he discovers things in his data base, interruptions for a sequence of batch job submissions can, I believe, be quite disruptive. Ideas can be lost and the sheer bother may discourage submitting a job to produce some graphs of marginal expected utility, --- when in fact one of these might harbor a major payoff. And one can demonstrate some interactive displays that would not be effectively reproduced in the passive mode. For instance, in interactive graphics, one can point to an individual case on a scattergram, or to a cluster of cases, and see how these map onto a set of simultaneously displayed scattergrams representing different variables. The nearest one might come to this in passive graphics is to print each case with a different symbol, which is impractical when a hundred or so cases are being plotted.

When thousands of cases must be examined, even a scattergram that is a cloud of points would be rather unreadable. In this case, some summary density map with a capability for zooming or for alternating displays between different data subsets is indicated, and again the advantages of interaction are apparent.

C. Tradeoffs between Form of Display and Method of Presentation

For some interactive systems that have a slow system for writing or for transmitting graphical information, or that have a limit on the complexity of what they can display at a given time, one probably will wish to adhere to simpler forms for data representation. For instance, because these systems usually approximate all curves by a sequence of straight lines, one may choose to represent the eye on a "face" in an interactive system by a diamond shape rather than by an ellipse, understanding that the latter would be quite acceptable in a passive graphics system. Actually, the diamond-shaped eye seems to be quite effective.

Furthermore, until the scientist becomes better acquainted with the more sophisticated data representation forms, he may progress more expeditiously with extended capabilities of familiar forms such as scattergrams and histograms. These familiar forms can take on considerable power in a responsive interactive system, and probably will require less time for his mind to decode.

D. Our General-purpose Programs for Exploratory Data Analysis

These programs are written for the IMLAC interactive graphics terminal accessing an IBM 360 or 370-series computer operating under TSO. Access can be by ordinary dial-up telephone lines, and we recommend modems that permit ≥ 1000 baud communication. Capabilities for coping with line noise, as well as devices for reducing the need to resend graphical data, are part of the basic IMGRAF (4,5) systems software, which also permits the user to write his programs in FORTRAN and hence to readily incorporate other FORTRAN-based software (e.g. the BMD statistical programs) within his programs (6). An applications-oriented software package, GRASP, provides the programmer subroutines that produce a variety of graphical displays, subroutines that manage entire interactive sequences such as menu selection and scrolling lists, and interactive subroutine-construction tools for commonly encountered needs.

We have developed, tested, and documented for export two interactive graphics programs for data exploration ( ). Another Imlac exploratory program developed by Michael Tarter (7) also should be noted.

<u>"SCATTERMAPS"</u> provides associated scattergrams for all of the cases in a data set. Data must be in rectangular-file form, up to 10 variables per case, with a name assigned to each variable. The program can accept a BMD Savefile as input, which facilitates the user's file-preparation task and access to BMD programs for further analyses. Up to four simultaneous scattergrams may be shown on the main display

frame. Beneath them are listed the various options the user may wish to select. He does so by pointing to an option with a light-pen, an optical device that instructs the computer that the display element it now "sees" is being chosen. This mode of interaction has two advantages: it avoids the delay and awkwardness of typing, and the user is not required to guess at what instructions exist or at what mnemonics may be used to represent them. At any time, for instance, the user may wish to change which variables are assigned to each graph. He points the light-pen to the words, "ADD A GRAPH", on the lower right-hand side of the main display frame (Figure 1), which then is replaced by a display frame that lists the names of the variables and permits the user to add a graph or select which variables he wishes to assign to the x and y axes of any scattergram, by pointing to them with the light-pen. When finished, he returns to the main display frame, which now displays the revised graphs.

Actions on the main display frame focus on intuiting multivariate relationships between individual cases by associating the points that represent them on a set of scattergrams, a simple step toward transcending the two-dimensionality of a scattergram. For instance, one may wish to see if a point that appears as an outlier on one scattergram stands apart from the main cloud of cases on the other scattergrams. The point is indicated by the light-pen, which causes points representing the same case on other scattergrams to be marked with a circle, and the values for all variables for that case to be printed at the bottom of the display frame

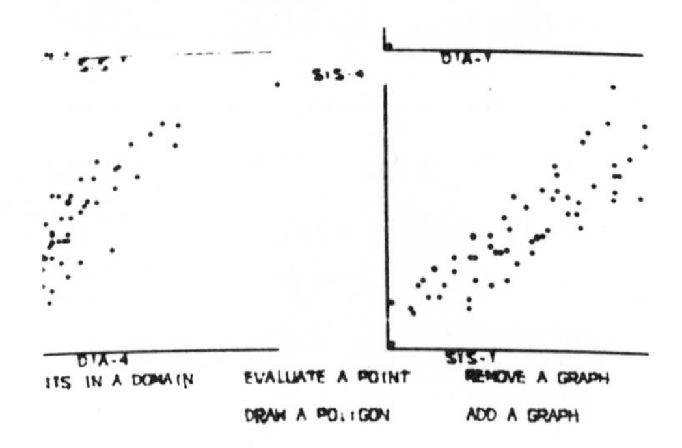

Figure 1: The user commands the system by pointing with the light-pen to instructions such as "ADD A GRAPH".

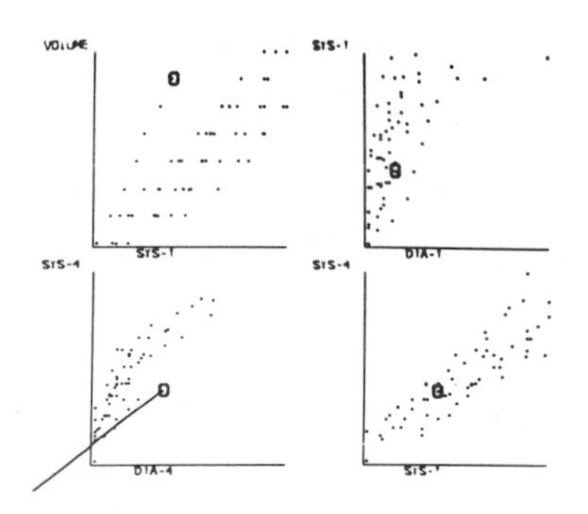

Figure 2: The user has designated a point on one graph by means of the light-pen. It and corresponding points on the other graphs then are identified by a circle, and the values of all variables for that case are as follows:

GROUP	=	0.100E 01	NUMBER	=	0.280E 02	Volume	=	0.300E 02
SYS-1	=	0.135E 03	D1A-1	=	0.900E 01	.DP-1	=	0.120E 04
-DP-1	=	0.800E 03	SYS-4	=	0.850E 02	D1A-4	=	0.280E 02
.DP-4	=	0.500E 03						

RD1

(Figure 2). This helps one decide whether this indeed may be an unusual case, or whether one of the variables has been erroneously transcribed or keypunched. Sometimes one's eye is struck by a sequence of points that appear to be closely associated along one curved line on a scattergram. If the user designates this sequence of points in order by the light-pen, they will be connected by straight lines in that sequence, on all of the graphs (Figure 3). This permits one to see if an ordering that was perceived with respect to two variables maps in some orderly way onto spaces defined by other variables. It sometimes is interesting to apply this to the points that bound one side of a scattergram. Finally, one wants to see how clusters of points on one graph map onto the others. The light-pen is used to designate a polygon based on points that bound the cluster. On the other graphs, cases that correspond to points on the polygon are marked with "+", and those within, with "0" (Figure 4). One thus can see not only if there is some clustering of these points on the other graphs, but also, even though the points be more widely scattered, if relationships within the cluster might be preserved (i.e. "0" points are for the most part enclosed within a perimeter of "+" points). Finally, one may want to see if relationships between sets of variables can be intuited by seeing how a domain that moves along a line in one scattergram maps onto the others. The domain is defined as for cluster examination, and the user then may designate the line along which it is to move. It moves in seven equally-spaced steps along the line, with the points contained by it being marked by "+" and "0" on all graphs as

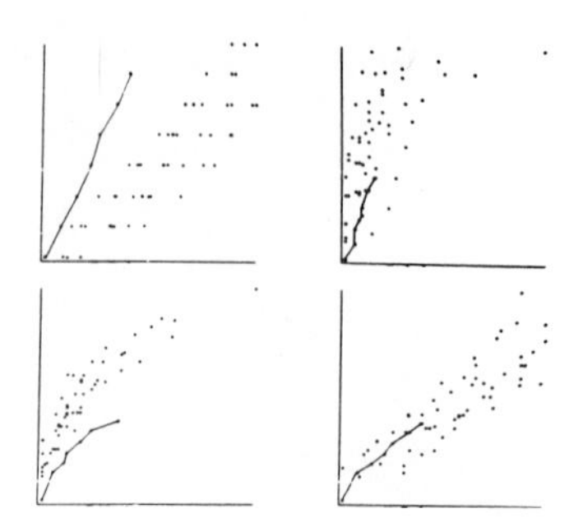

Figure 3: The user has designated a sequence of points on one graph. On it and on the other graphs, the sequence is connected by straight lines in the order in which the points were designated.

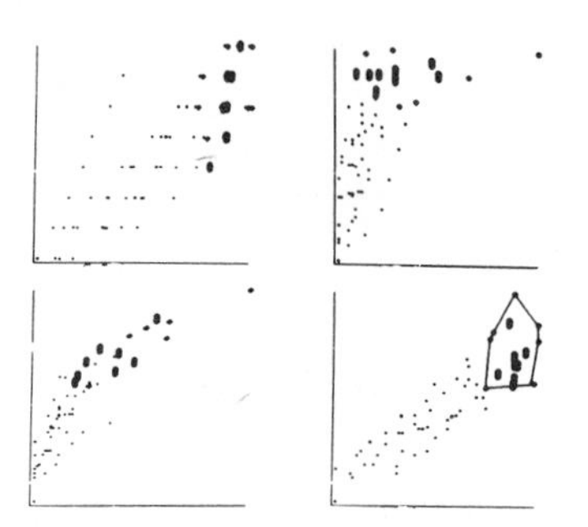

Figure 4: The user has designated a polygon surrounding a cluster of points on one graph. On all graphs, points on the polygon boundary are marked with "+", and those within the polygon are marked with "0".

described above. The user may control the speed at which the sequence of displays is shown, opting for closer examination of a slowly stepped sequence, or a more dynamic sequence that instructs through a sense of motion. For instance, with the faster dynamic display, frames that correlate poorly can be readily identified by the random scintillation of "0" marks on them. On other frames, one can get a sense of trajectory for the points, -- e.g. fairly stationary clumping for the first few frames followed by a rapid diagonal progression of the clump for the remainder. Obviously, a capability for transgeneration should be added.

Two additional display frames permit relationships to be visualized dynamically in three-dimensional space. The user specifies which variable is to be the new z dimension, and the display is rotated about the present y direction through a centrally located origin. One display frame may be requested when a sequence of points has been connected by lines. The lines are rotated to see if similar or other interesting relationships hold in three dimensions. In another display frame, any one of the scattergrams may be selected for rotation of all of its points. In both cases, the user may control the speed of the rotation.

Finally, one scattergram may be selected for display in association with a histogram representing a third variable. Its points then are plotted as numbers that cross-reference to the histogram. With the light-pen, one may indicate a point on the scattergram, which then is cross-referenced by a line to its exact position on the histogram, values of the three variables being printed out to the side (Figure 5).

One may ask that scattergram points related to a bar in the histogram be enhanced (marked by large asterisk) for easier detection among the others. One also may ask that points on the scattergram be plotted so as to reflect the passage of a specified window across the histogram: By light-pen, one indicates the width of the window. Thereafter, all points on the scattergram that fall outside of the window are plotted as dots, and points within the window are plotted as "X"s. As the window moves, different subsets of the points are thus enhanced. Again, do we see some order in the movement, or do the "X"s scintillate randomly?

We all see things differently. For some people, static displays are quite adequate. For others, such as the writer, an added intuitive grasp of relationships is imparted through observing motion as a series of frames is displayed sequentially in the same position. There is a superposability here that differs somewhat from what one can visualize if a row of similar separate graphs is shown on a static display. Also, the screen for the single dynamic graph is greater than that which would be available to any one of the graphs in a static, non-superimposed sequence on the same screen. The decision to opt for dynamic displays influences one's choice of an interactive graphics system, as will be discussed in a later section.

"CLASS COMPARISONS" enables classes to be compared with each other, and individuals with classes. For instance, one might want to see which diagnostic classes a patient appears to belong to, and how he differs from the ideal in each. Or one might want to compare how patients assigned to different

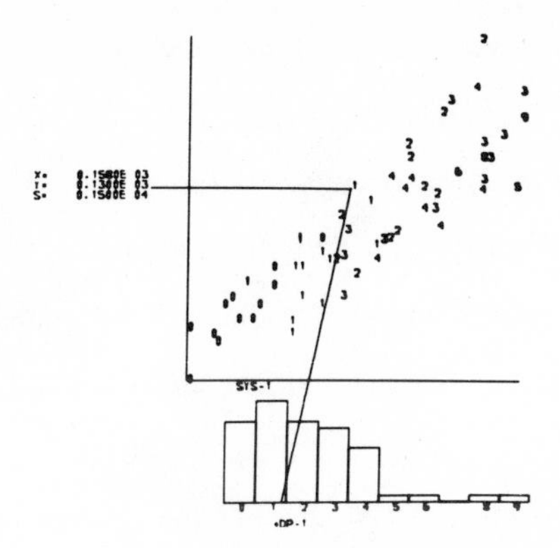

Figure 5: Numbers constituting a scattergram for two variables refer to locations on the histogram representing a third variable. The user has just indicated a point on the scattergram, and the computer has responded by designating its exact location on the histogram and printing the values of the three variables for it.

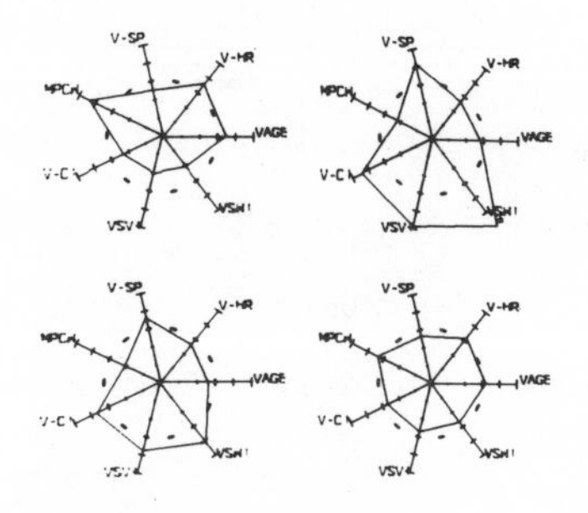

Figure 6: This is the main display frame for the "Class Comparisons" program. Data are from patients being followed in a cardiology clinic. See text for explanation of the polygons. Similarity of shape for polygons for "DEAD" and "KILLIP 4" patients suggests that Killip 4 is not a favorable clinical classification.

classes of treatments have progressed after a certain amount of time. The data base is visualized as one set of scalar variables and one set of classification variables. The scalar variables would include measures such as systolic blood pressure, yearly cost of health care, interorbital distance, etc. Up to 24 of these can be accommodated by the program at this time, but no more than 20 can be represented on the displays (10 for "faces") at any time. Up to 20 types of classes can be specified, --- e.g. status on discharge, sex, status with respect to whether any of a certain class of operations has been performed, type of health-care delivery system used. Under each class type, the case may be assigned a positive integer number (up to 20) indicating its status (e.g. 3 might designate "private group practice" under category "type of health-care delivery system used"). The user may indicate a common "missing data" code for all scalar variables, or specify separately for each variable. It may be a single number, or it may be a lower or upper bound. There is no limit on the number of cases that may be handled, except for the capacity of one's disk file. A manual with sample output guides the new user through all of the options he might wish to use, though self-instruction with some trial and error should not be very difficult without the manual. There also are instructions on how data are to be prepared for the program. BMD Savefile cannot be used because of its lack of provision for naming sub-categories, -- i.e. names associated with different numbers under each classification category, such as "private group practice" with "3" in an example cited above.

Figure 6 is an example of the main display frame. A "test case" or "test class" may be compared with up to four reference classes. Under each polygon is the name of the reference class it represents, as well as the number of cases in that class. The name of the "test" individual or class being compared with these is shown at the top of the display. At the bottom of the display are the various options to which one can branch from this display frame, pointing to the desired one with the light-pen.

Consider one of the polygons (Figure 7). Each arm is labelled with a 4-character reminder of the scalar variable it represents. One can get a more complete description of each of these by pointing to "EXPLAIN A LABEL". Notice the marks on each arm. The third from the center is on a unit circle which is characterized by dashes between the arms. If the average value for the variable represented by this arm is the same for both "test" and "reference" classes, the polygon vertex will be plotted at this unit-circle position. If the average value for the "test" class is one standard-deviation greater than that for the "reference" class, the vertex will be plotted at the next mark away from the center (the 4th mark out). Standard deviations are based on the reference classes.

The user indicates the variables he wishes to assign to the polygon arms by pointing the light-pen to items on the list of names of variables that will be displayed at the outset of the program or after he points to "CHANGE VARIABLES DISPLAYED". At any time, he may exchange the assignment of variables to the various arms or may replace the

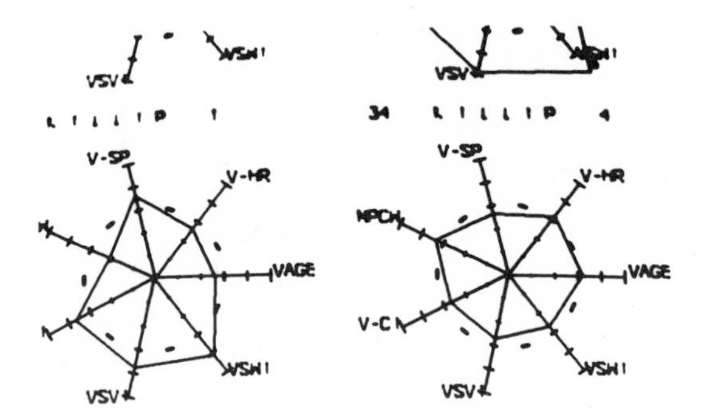

Figure 7: Closeup of a polygon in "Class Comparisons" program. See text for explanation.

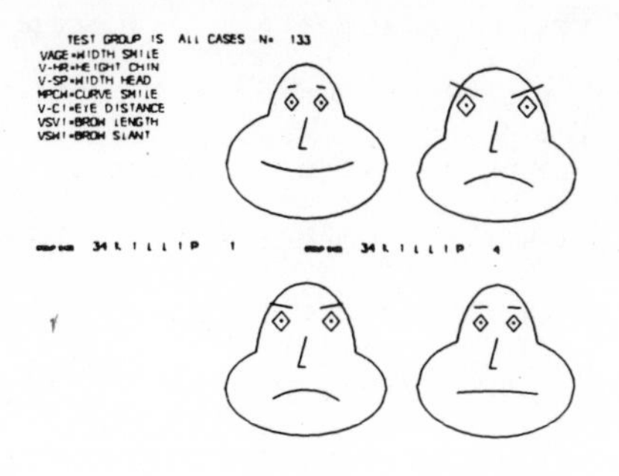

(a)

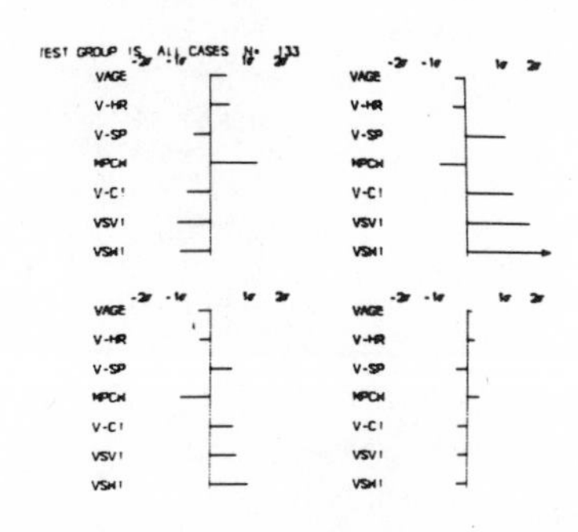

(b)

Figure 8: Alternative representations may be requested for the same data, (A) faces, (B) bar graphs.

general form of the representation, choosing among polygons, faces (Figure 8a), and a simple bar display (Figure 8b). He thus may readily interchange assignment of variables to facial features, removing concerns about biased presentations and enabling some sense of how dependent visual classifications might be to these assignments. Whereas faces enable people to recognize class similarities rather easily, it is helpful to convert to polygons or bars to better quantitate some of the differences. As one moves from bars, through polygons, to faces, one has stronger cues from shape and decreasing quantitative grasp (less so as one goes from bars to polygons). Most people we have encountered, once other forms lose their novelty, seem to settle for a compromise at the polygon level. However, there probably is an advantage in being able to move freely from one form to another.

"Reference" and "test" classes may be selected as subsets of the data base, or "test" classes or individuals may instead have their scalar data entered from the console, e.g. one might want to study a new patient with respect to others in the file. For subset selection, the user is presented a list of the various types of category, from which he may make a light-pen selection of one (e.g. "PAST MYOCARDIAL INFARCTIONS"), which brings up a list of the entries available under this category (e.g. "NONE", "ONE", "TWO", "MORE THAN TWO"). From the latter list, the user may select as many items as he wishes, and all cases will be included that belong to any of the selected subcategories. The user types in the name of the new class as he wishes it to appear on the main display frame.

In the example shown here, it can be seen that people classified as KILLIP I resemble people who die far less than do those classified as KILLIP IV. In other display frames, one can see that people with a past myocardial infarction or hypertension do not differ much from the average patient in the entire group (all cardiology patients), but that patients with rales have a more ominous pattern (Figure 9). Most important, one can form opinions about the variables that seem to differ most from one class to another. Clearly, other formal methods can detect relationships such as these, but one suspects that their outputs might not be as readily assimilated as are the graphical representations while a scientist is rapidly exploring his data base during a discovery process.

One dynamic display is available in this program. The user may select any variable, and all cases are sorted with respect to this variable. The range of the variable is then divided into 10 equal segments, and a separate display is generated for cases in each of these groups. These displays are then sent over to the terminal, which displays them in a sequence whose speed can be controlled by the user, as in the SCATTERMAP program. This is a minimal display, -- e.g. the polygon only, with arms and labels (except for one display) removed. Such is necessitated by the fact that all of these displays must be stored in the terminal's memory before they can be displayed in rapid sequence. This memory is limited, so complex displays such as "faces" are limited with respect to the number of frames that can be accommodated.

Again, the user may wish to examine a slower sequence of frames more minutely, or he may form a useful qualitative appraisal from a rapid display. For example, suppose we sort on age. If the polygon merely quivers randomly as the dynamic sequence is displayed, the multivariate relationship between the "test" and "reference" groups probably is rather age independent. For the cardiological data mentioned above, this seems to be the case for all but the youngest and oldest categories.

SCATTERMAPS and CLASS COMPARISONS complement each other rather well, and it probably would be worthwhile to develop software that enables them to be used interchangeably at one investigative session. When the number of cases is large, one cannot visualize all data well on a scattergram. The selection of interesting sub-categories for scattergram display might be guided by the CLASS COMPARISONS program. On the other hand, clusters or individuals identified on the scattergrams might be worth investigating with respect to various reference classes.

E. Some Special-Purpose Programs

Rectangular tables fall far short of being adequate coordinate systems for many naturally occurring data sets. How much more meaningful it is to study relationships between heart disease data and demographic data in reference to a map of the various county districts with which one is familiar. Data groupings may bring to the experienced mind some postulated relationships, not presently in the data base, that are worthy of further study; e.g. if one were to note a northwesterly path of high incidence of respiratory disease, diverting appropriately around the bases of some hills, one might look for some factories at its origin and study contaminants in the fumes that they release.

An excellent example of interaction and of taking advantage of natural structures was programmed by a graduate student in neuropharmacology (8). He had observed both behavioral and physiological outcomes when various pharmacological agents were injected into various sites of a cat's brain, with or without pre-conditioning medications. After many experiments on many animals, he found himself with a mound of data that were nearly impossible to analyze, especially with respect to their relationships to underlying pathways and other structures in the brain. The object was to find not only which drugs might induce which outcomes, but also the brain sites at which they are most active. Furthermore, in active animals, any post-injection observations would include both random and related outcomes. He placed a three-dimensional brain atlas in the computer, a sequence of diagrammatic coronal sections, together with his experimental data on treatments and outcomes. The drug, method of injection, and premedication for an investigation were rapidly selected by the light-pen from lists ("menus"), and the subfile these defined was extracted from the main file. Then the student selected any Boolean combination of outcomes he wanted to study. On the brain-map display, all injected sites in that section were marked, with a "+" for all sites where the specified outcome was seen and a "-" for all sites where it was not. These symbols could be converted to numerical ratios, e.g. "2/5" for a site where 2 positive outcomes had been observed for 5 injections. On a small sagittal section of the brain, orthogonal to the coronal sections, a line indicated the site at which the currently displayed large coronal section was located. The student could skip from one coronal section to another by pointing with the light-pen to the appropriate place on the sagittal section diagram, or by calling up a sequence of displays in either direction by pointing to the words "ANTERIOR" or "POSTERIOR". At any time he wished, he could study a new treatment or change the Boolean outcome. Needless to say, this expedited his initial search for hypotheses in his data base immensely. He claims that it was well worth the programming effort for this one study alone.

Whereas graphics may have a great advantage for special cases such as these, it must be conceded that better software to aid special program development probably is needed to break even in effort for typical individual studies. Our GRASP subroutines can help considerably, as can applications-oriented subroutines in other studies, but further study of typical requirements of special systems is desirable to add what is needed to substantially lower the barriers to special program development. To this end, we have encouraged special program development by a variety of users on our system, observing where they need help most and translating these findings into GRASP and IMGRAF software, as time and budget have permitted.

III. THE OMEGA

When scientific findings are applied to complex problems in the practical world, very often a new group of people become involved, the policy and decision makers. It often may be helpful to instruct them in relationships that have been established, with the aid of displays that first helped the scientist to intuit these relationships. But now it is an ethical necessity to advise the new learners that displays such as these only help to make clearer relationships that may be perceived in one's data, -- that considerable research and hypothesis validation is required to narrow down to those relationships that are not accidental or artifactual. They should be clearly advised whether the displays being shown them illustrate well-established relationships or whether they are purely exploratory, with all of the risks that are entailed. There is an honest difference of opinion on this point, but I believe that it is far more important to be very scrupulous with respect to these kinds of exhortations than to avoid certain less objective graphical forms, such as the faces, particularly if the user is given and is urged to use an ability to vary the display (e.g. reassign variables to features).

I shall mention only briefly one aspect of interfacing policy and decision makers that goes beyond the scope of this conference, but that should nevertheless be in mind. I believe that data analysis systems should be associated with exploratory modeling systems into which it then becomes easy to introduce findings that have been established. Using models representing established premises, or capable of being rerun comparatively over the plausible range of premises not yet well established, policy and decision makers should be able to play through the consequences of imposing upon the models the various constraints and inputs that characterize the alternative policies or decisions that are being studied. It is a model embedded in another, manipulative, model. My experience in modeling strategies for treating cancer leads me to believe that the same basic interactive graphics system that can place the scientific investigator in productive contact with his data base also is an unusually effective vehicle for exploring the models entailed in studying strategies for complex applications, given the appropriate applied software.

IV. IN BETWEEN

This area is covered rather well by formal methods that have been developed over the years and continue to advance. Also, the combined process of experimentation and analysis is rather protracted, placing at lesser advantage the qualities of expeditiousness and immediate flexibility that are valuable for initial data exploration. Nevertheless, we have found that there are some advantages in good interactive graphics presentations of methods such as regression, discriminant function analysis, factor analysis, and time-series analysis. Most generally important is the ability to rapidly reinitiate a computation with revisions made on the basis of observing effective graphical and numerical displays. For instance, one may want to revise a filter, test effects of deleting a data point, guide a rotation, or revise the representation of a function for regression, e.g. from $Ae^{-Bt} + Ce^{-Dt}$ to $Ae^{-Bt}(1 + Ee^{-Ft})$ if B and D prove to be close.

Our factor analysis and time-series programs have been primarily investigative, but the discriminant function and non-linear regression programs have been refined and rather well tested for use by others. Both make use of BMD software for the basic statistical computations.

The present non-linear regression program is designed to help biologists relate to data the rather simple bivariate models they often encounter, -- e.g. growth curves, sums of exponentials. It could be extended to additional independent variables with very little effort. The well-tested version of this program requires the user to specify both the function to be fitted and its partial derivatives at the console, an interpreter having been written in FORTRAN to handle these. A graduate student has recently written a version, based on a more recent BMD-Q program, that does not require specification of the derivatives. When well tested and released, this should attract more biomedical users. We have found that biologists tend to lack experience in dealing with functions, which makes it difficult for them to obtain reasonable estimates of initial parameter values. Although the BMD programs are designed to minimize the effects of this handicap, they can't do away with them in all cases. The interactive graphics program enables the user to revise parameter estimates, observing the effect on the curve representing the function, which is plotted in conjunction with the data. He thus has visual guidance as he works toward reasonably good initial values,

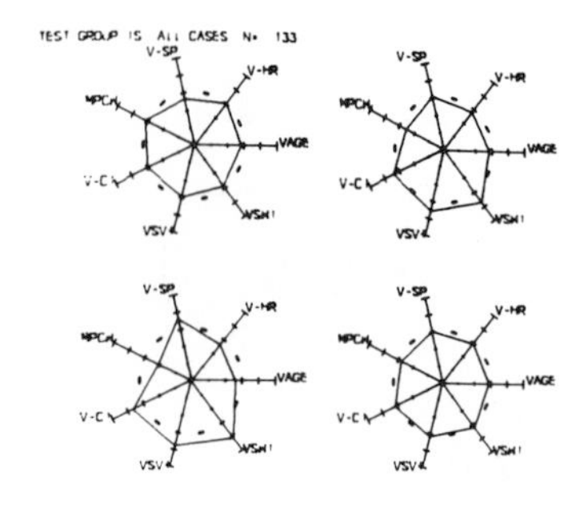

Figure 9: The nearly circular polygons for patients with hypertension (upper left) or one myocardial infarction (lower right) suggest that they differ little from the average of patients in that clinic. The shape of the polygon for patients with rales (upper right) is somewhat more ominous.

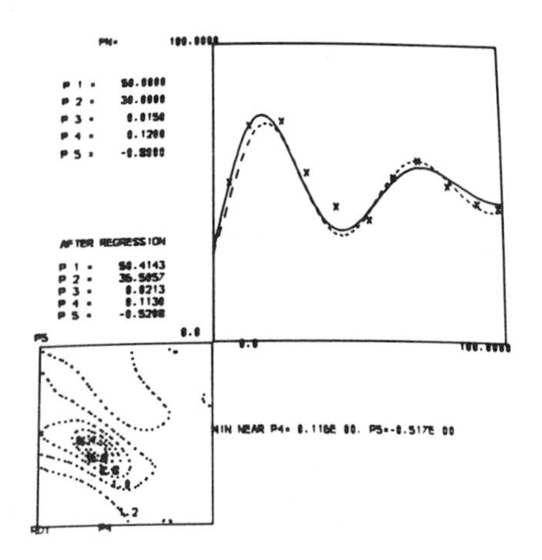

Figure 10: User may vary initial parameter estimates of a function until its graph (dotted line) appears to approximate the data fairly well. Then a non-linear regression calculation produces the solid curve and the lower set of parameters on the left. A contour plot of summed squares of residuals over designated ranges of two variables is seen at the lower left.

and this proves to be an effective teaching aid for his future management of functions. After the regression has been completed, regression parameters and a solid plotted curve can be compared with the user's parameter estimates and their corresponding dashed curve. Adequacy of the model can be assessed with the aid of a plot of residuals. The user may also request a contour plot of summed squares of residuals over specified ranges of any two parameters (Figure 10). This helps assessments of whether the solution is at a relative or absolute minimum, and of why a computation may not converge readily (e.g. where there are two close exponentials).

V. THE ENTIRE DATA ANALYSIS PROCESS

Experience has taught that it is quite possible to fragment this process into a sequence of separate batch jobs, though it is useful to be able to modify and hand on a data file (e.g. BMD Savefile) that can communicate with the various programs that are to be used. Many of the people who operate this way would join in characterizing data analysis ideally as a process having continuity and the ability to recycle at many levels. After gross cleanup, data can be further screened by programs that have been developed for exploratory data analysis. As one views one's data, the need for derived variables is seen, and one should be able to generate these as one proceeds. Data subsets for more formal analysis should be readily selectable from the underlying files by the usual retrieval methods, as well as by visual definition of interesting clusters during use of the exploratory programs. Formal analyses yield the parameters for additional derived variables, which then can be visualized in the exploratory programs as well as in visual displays associated with the analytical programs. Versatile file management, free movement from one mode of analysis to another, and effective graphical displays would seem to be essential components of a good complete data-analysis system. Some statistical systems now provide certain aspects of these components, but the writer is unaware of systems that provide them fully in the context of a user-oriented interactive graphics system that extends to all levels of data analysis and includes versatile file management. The writer has been studying design criteria for such a system and believes that a user-oriented interactive-graphics implementation is virtually essential, especially if scientific investigators as well as statisticians are to use it.

VI. OBSTACLES

If this all is so great, why isn't it in more widespread use? The most important reason probably is that most of this type of work so far, --- e.g. at UCLA's Health Sciences Computing Facility, Stanford (PRIM-9), and the National Bureau for Economic Research has been primarily investigative. A commitment to document, export, and maintain programs entails a major effort and requires corresponding additional financial support.

Another problem is limited portability of graphics programs. Even batch programs written in FORTRAN require checking and often some modifications when transferred to other computers; the transfer of interactive graphics programs is far more difficult. Interactive systems must interface to a computer's operating system, which may vary from one installation to another even though the basic computer is the same. Furthermore, despite major efforts of ACM's Special Interest Group on Graphics, as well as others, to negotiate agreements on graphics standards, there is imperfect standardization today with respect to software interfaces to different graphics devices. There is considerable need to write basic graphics software systems, as well as applied programs, in such a way that the development of interfaces to new graphics devices or host processors becomes a limited, well-defined task.

Expense has been regarded as another important problem. This probably is a realistic assessment for an individual user who is engaged in relatively routine problems. However, for the types of complex undertakings for which an interactive graphics approach is eminently suited, the expense of acquiring an interactive graphics system should be assessed in comparison with expenses entailed in acquiring the data base and in salaries to investigators whose work may be greatly expedited by this approach. A quite useful Imlac configuration, complete with light-pen, long-vector hardware, and a communications interface costs something over $20,000. Basic storage-tube systems are much cheaper unless they are configured as intelligent terminals, which introduces a processor that represents much of the difference in cost between them and the Imlac, which is an intelligent terminal. Prices are bound to fall and capabilities (e.g. color, grey levels) increase as raster graphics enter the market.

A major obstacle probably could be to persuade people with established habits of performing analyses in "rapid batch" to try something new. Perhaps they should be bypassed, with an emphasis instead on introducing scientific investigators and students in statistics to the new interactive approaches, along with the more conventional approaches. Observe and learn from what they choose in the long run. We have derived valuable insights from some of our resistant statisticians. For instance, they point to their need to flip back and forth rapidly between pages in their output, and to view large tables that would not fit on any but possibly the largest storage-tube screen. A good interactive graphics system should be able to meet these needs. It can, for instance, improve on the page flipping by making it easy for the user to juxtapose on one screen the graphs or tables between which he is flipping in his printed output. Very likely the appropriate graphs or tables could be located more rapidly by interactive retrieval than by his search through a thick stack of output. Also, what is he seeking when he looks at a huge table? He can't focus on all of those numbers, and hence must be forming impressions of magnitudes with more precise visualization of numbers in regions where the magnitudes are of interest. This can be approximated, but perhaps not much improved on, by interactive displays that depict magnitudes and then zoom to more detailed numbers in regions the user indicates with a light-pen. In all of these remedies, we are assuming that the graphics display can be changed rather rapidly. Only systems that permit this should be seriously considered for data analysis.

VII. CHOICE OF A SYSTEM

Available software is the most important consideration, both appropriate applied programs and systems that facilitate the construction of new applied programs. Software development can become a major cost center otherwise.

If a variety of users or data suppliers are to have access to a central system, and if their work would be expedited by their having interactive graphics terminals, a distributed system must be considered. One therefore wants a terminal for which communications software has been well developed. Don't look only at the mere ability to communicate. Consider as well the capabilities for handling noise and for minimizing the resending of display data. The intelligent terminals one would want for such a system should also be evaluated for their stand-alone capabilities. Considerable data editing and screening could be done locally before data are sent on to the central system.

If dynamic displays or light-pens are desired, refreshable scopes are to be preferred to storage scopes. However, the latter can accommodate displays of almost unlimited complexity, whereas the refreshable scope has limits on what it can display and may flicker as these limits are approached. Refreshable scopes, having higher contrast, are easier to view in well-lit rooms than are storage scopes.

Needless to say, the system must be able to run stand-alone or on the computers and operating systems that are at one's disposal.

REFERENCES

1. Chernoff, H., (1972) "The Use of Faces to Represent Points in n-Dimensional Space Graphically," Proc. of the Computer Science and Statistics Sixth Annual Symposium on the Interface, October 16-17, University of California at Berkeley, M. E. Tarter, Editor, pp. 43-44.
2. Friedman, H. P., Farrel, E. J., Goldwyn, R. M., Miller, M. and Siegel, J. H., (1972) "A Graphic Way of Describing Changing Multivariate Patterns," ibid., pp. 56-59.
3. Newton, C., (1976) "Graphic Data Analysis: Optimizing Tradeoffs Between Richness and Simplicity," Proc. 1976 Statistical Computing Section, American Statistical Association, pp. 238-240.
4. Ryden, K. H. and Newton, C. M., (1972) "Graphics Software for Remote Terminals and their Use in Radiation Treatment Planning," Proc. Spring Joint Computer Conference, 40:1145.
5. Newton, Carol M. and Ryden, Karl H., (1972) "Remote Interactive Graphics and Cancer," First USA-Japan Computer Conference Proceedings, October 3-5, 1972, Tokyo, Japan, pp. 277-286.
6. Yuen, K. K., (1972) "A Graphics Spree for BMD," Proc. of the Computer Science and Statistics Sixth Annual Symposium on the Interface, ibid, pp. 52-55.
7. Tarter, Michael, (1978) "Implementation of Harmonic Data Analysis Procedures," Proc. of Computer Science and Statistics Eleventh Annual Symposium on the Interface, North Carolina State University, Raleigh, N.C., March 1978. Edited by A. R. Gallant and T. M. Gerig, pp. 234-239.
8. Sheu, Y., et al.,(1969) "Topographic Information Retrieval in Neuropharmacology by Using Graphic Display," Proc. of 24th Annual Conference of the Assoc. Comput. Mach., pp. 483-489.

ON CHERNOFF FACES

Lawrence A. Bruckner*

University of California
Los Alamos Scientific Laboratory
Los Alamos, New Mexico

Herman Chernoff introduced the idea of using faces to represent multidimensional data in 1971. Since then, this technique has been used in a wide variety of applications.

The first part of this paper discusses how to use the technique. Then Andrews' sine curves and Anderson's metroglyphys are introduced and compared to the facial representations. Dependencies among the facial features are considered next and a way to eliminate dependencies presented. Finally, some uses of Chernoff Faces at the Los Alamos Scientific Laboratory are mentioned.

I. INTRODUCTION

The use of Chernoff faces to investigate multidimensional data has been accelerating in the last few years. The most common usages of the technique are to display the data in a convenient form, to aid in discovering clusters and outliers, and to show changes with time.

The idea of using faces to represent multidimensional data was introduced by Professor Herman Chernoff under a contract with the Office of Naval Research while at Stanford University in 1971 (1). Professor Chernoff considered data having a maximum of 18-dimensions and allowed each dimension to be represented by one of 18 facial features. A typical Chernoff face is presented in Fig. 1. Herbert T. Davis, Jr., added nose width and ears to the face while at the Los Alamos Scientific Laboratory (LASL) in 1975. This revised face is shown in Fig. 2.

---

*This work was supported by the Conservation Division of the U. S. Geological Survey, Denver, Co.

ISBN 0-12-734750-X

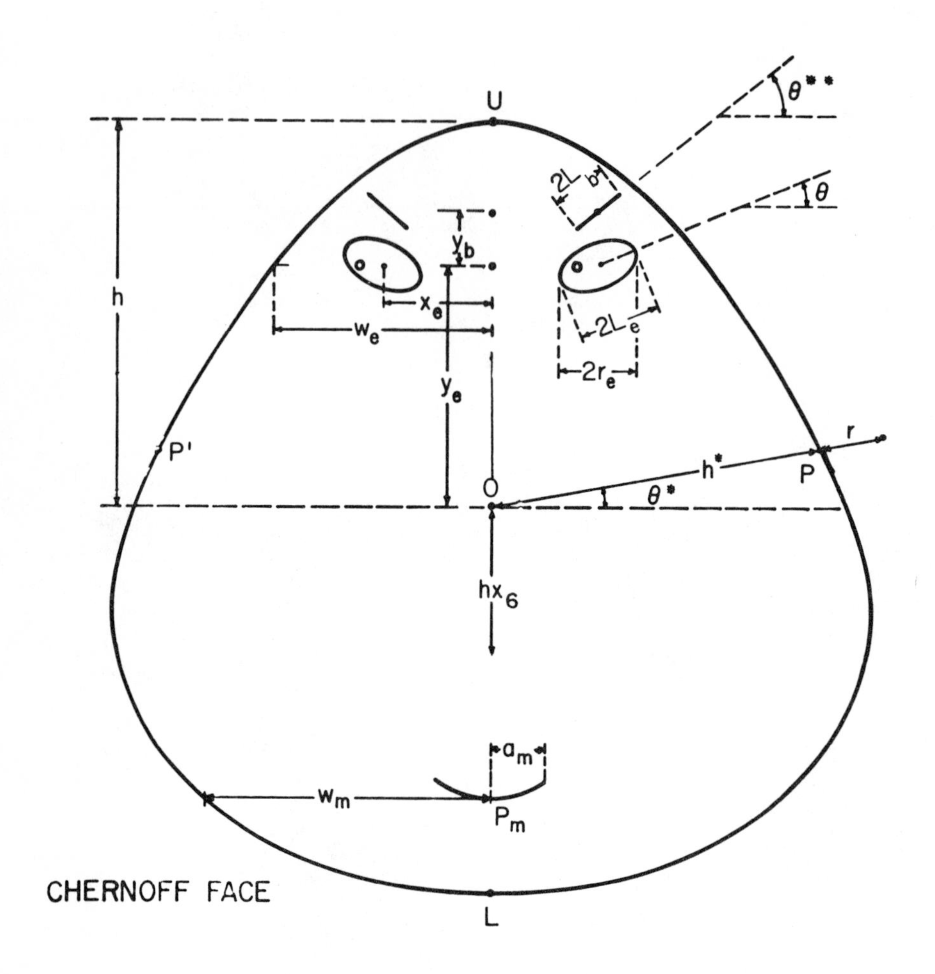

Figure 1. The Original Chernoff Face

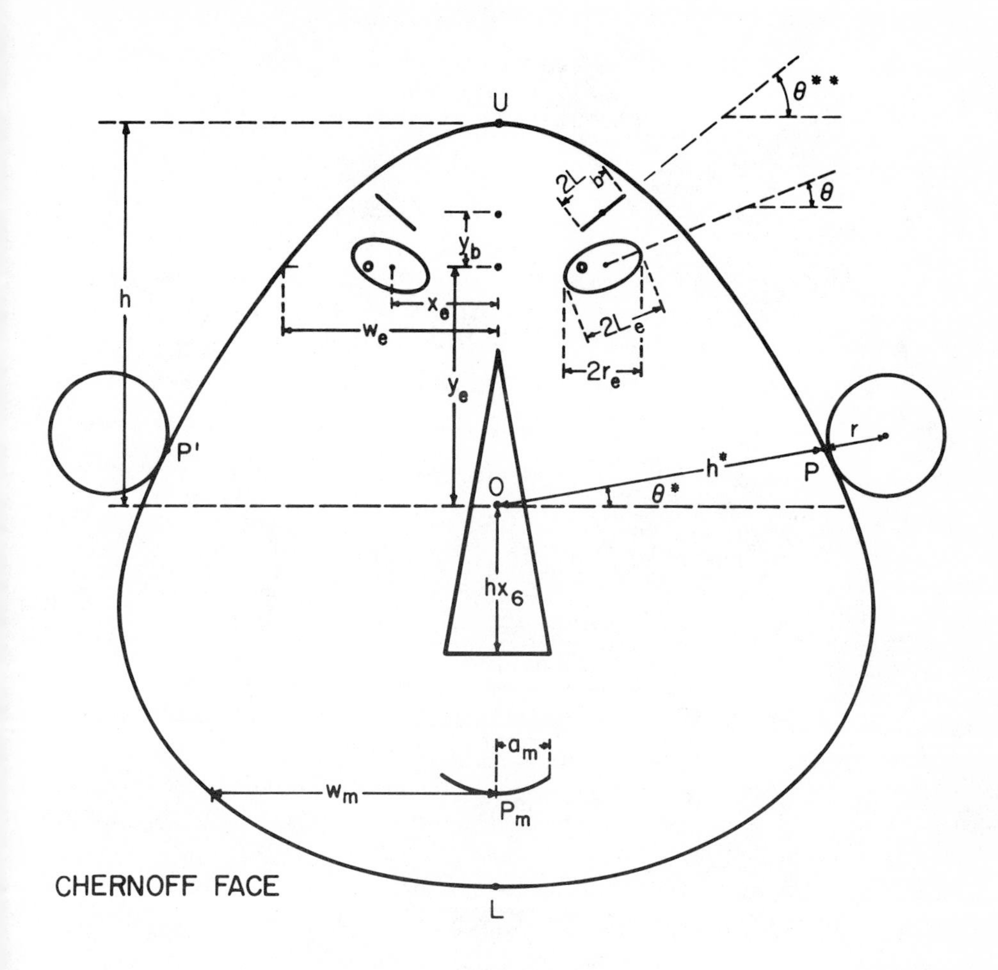

Figure 2. Davis' Chernoff Face

Table I identifies the facial features, the range of values each may assume, and the value the program assumes for the facial feature when that feature is not used to represent a data dimension.

TABLE I. Description of Facial Features and Ranges

Variable		Facial Feature	Default Value	Range	
x_1	controls h*	face width	.60	.20	.70
x_2	controls θ*	ear level	.50	.35	.65
x_3	controls h	half-face height	.50	.50	1.00
x_4	is	eccentricity of upper ellipse of face	.50	.50	1.00
x_5	is	eccentricity of lower ellipse of face	1.00	.50	1.00
x_6	controls	length of nose	.25	.15	.40
x_7	controls p_m	position of center of mouth	.50	.20	.40
x_8	controls	curvature of mouth	0.00	4.00	4.00
x_9	controls	length of mouth	.50	.30	1.00
x_{10}	controls y_e	height of center of eyes	.10	0.00	.30
x_{11}	controls x_e	separation of of eyes	.70	.30	.80
x_{12}	controls θ	slant of eyes	.50	.20	.60
x_{13}	is	eccentricity of eyes	.60	.40	.80
x_{14}	controls L_e	half-length of eye	.50	.20	1.00
x_{15}	controls	position of pupils	.50	.20	.80
x_{16}	controls y_b	height of eyebrow	.80	.60	1.00
x_{17}	controls θ** −θ	angle of brow	.50	.00	1.00
x_{18}	controls	length of brow	.50	.30	1.00
x_{19}	controls r	radius of ear	.50	.10	1.00
x_{20}	controls	nose width	.10	.10	.20

Few of the facial descriptions are entirely accurate. Most of the facial features are controlled by the data associated with the feature and the data associated with other features. For example, the true face width is a function not only of h* but also of θ*; mouth length depends on a_m and also on w_m.

The ranges of the facial features have been adjusted so that the faces look more "human" and so that all the features are observable. The eye size has been set so that the pupils can be seen; the mouth length set so that curvature is visible. It is important that all features be observable; that the faces possess human-like features is a matter of preference and appropriateness. It may be that the use of human-like features will contribute to the interpretation of one set of data but not to another.

II. USING THE PROGRAM

To create a Chernoff face an assignment of the data dimensions to the facial features is made. This assignment may be made at random or deliberately. Some users prefer the random assignment to reduce subjective elements, others deliberately employ perception of facial characteristics in the assignment. Thus, a measure of success or failure may be associated with mouth curvature and a measure of liberal/conservative stance with pupil position (looking to the left or right).

Once the assignment is made, low and high data values of each data dimension are determined. The actual value of the data variable will be linearly mapped from the data range into the facial feature range. (It is sometimes advantageous to transform the data by logs or powers before carrying out this step.) When all data ranges are set, the data can be mapped into facial feature ranges and a face which represents the input data drawn.

The data ranges should be set carefully. If the range is set too small, it will not include all of the actual data values; if it is set too large the data values may be too close relative to the range and a loss of discrimination in the facial feature may result. Figure 3 shows the results of data values outside of the mouth length and nose length ranges. Transformed data may spread the actual values more or less uniformly throughout the range and this can lead to increased differentiation. Only knowledge of the data can help one decide what to do.

A listing of the control cards, the computer program, DRFACE, and the data from a study on oil companies involved in outer continental shelf leasing and drilling is contained in the Appendix. Note that the READ statements for the facial feature - data variable assignment allow easy reassignment. This is particularly helpful when the data are to be viewed from several perspectives.

The user specifies one input and two output format statements. The two output formats allow neater display. The actual output for the oil company example follows the data in the Appendix. The facial features chosen, their ranges, the corresponding data variables and their user-given ranges are printed first. Then a three-line set of information for each face drawn is printed. The first line gives the scaled facial data and the third line gives the number of the corresponding facial feature.

From the output data we see that the facial feature "mouth curvature" is associated with the data variable "Royalty per Production Year." (See Table IV for a definition of the latter term.) The low and high values of "Royalty per Production Year" are assigned on the input cards; the low and high values of "mouth curvature" are assigned in DRFACE in the DATA FEAT statement. Figure 4 contains the faces drawn for this example.

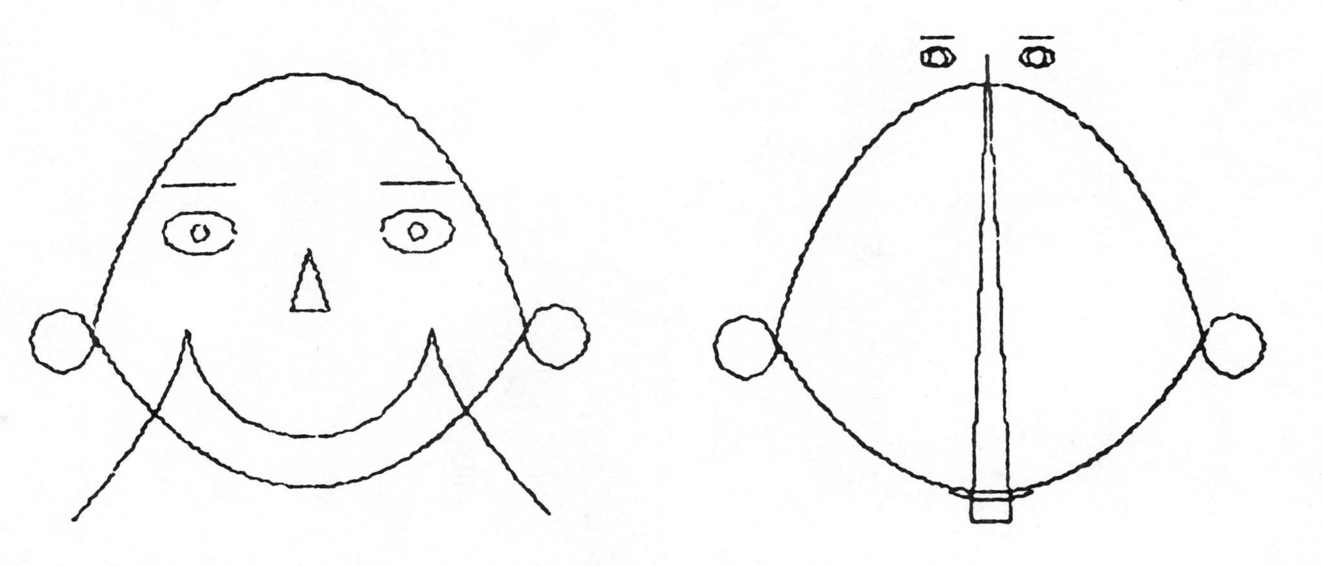

Figure 3. Effect of mouth length (a) and nose length (b) being out of range

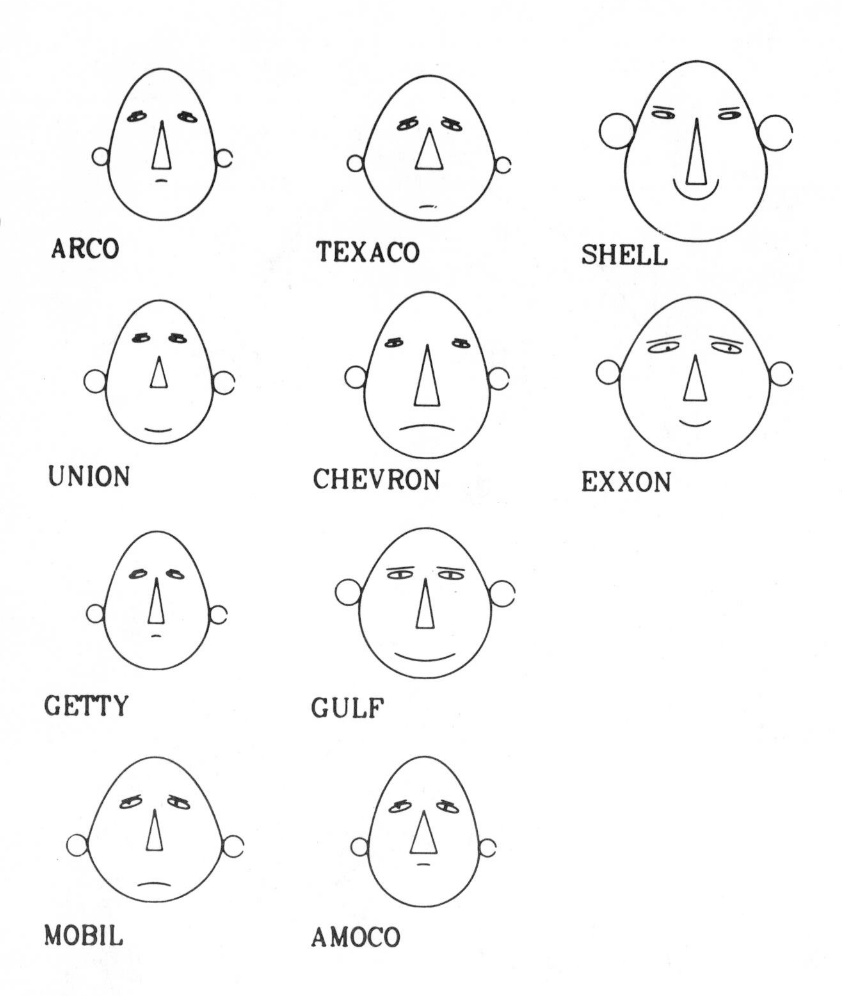

Figure 4. Chernoff Faces for 10 major oil company groups

At LASL DRFACE is run on both the CDC 7600 and CDC 6600 computers. The former version is the program given in the Appendix. The faces are produced on a Calcomp model 565 plotter. The 6600 version is used in conjunction with a Tektronix 4000 Series CRT screen terminal and a film recorder.

III. OTHER TECHNIQUES FOR DISPLAYING MULTIVARIATE DATA

There are many ways to display multivariate data. Table II presents the data for the oil company example. (The variables are described in Table IV). A look at this table conveys very little information. Figure 4 in the previous section presents the data by means of Chernoff faces.

The same data will now be displayed using Andrews' sine curves and figures called metroglyphs. D. F. Andrews (2) has suggested mapping multidimensional data into trignometric functions on $[-\pi,\pi]$ in the following way

$$(x_1, x_2, \ldots, x_k) \longrightarrow \frac{1}{\sqrt{2}} x_1 + x_2 \sin(t) + x_3 \cos(t) + x_4 \sin(2t) + x_5 \cos(2t) + \ldots .$$

This function is then plotted so that each multidimensional point produces a curve. The curves are viewed and those that lie close together represent clusters. The results for the oil company data are presented in Fig. 5. I find this figure hard to interpret. Shell appears to be different, but the rest are too intertwined. Sometimes plotting principal components rather than the data improves the picture. It did not help in this case.

The first seven variables of the data in Table II are plotted in Fig. 6. These shapes are called metroglyphs (3). These are typical of many other types of multidimensional data display techniques which use circles, rays and location within an area to display the data. This

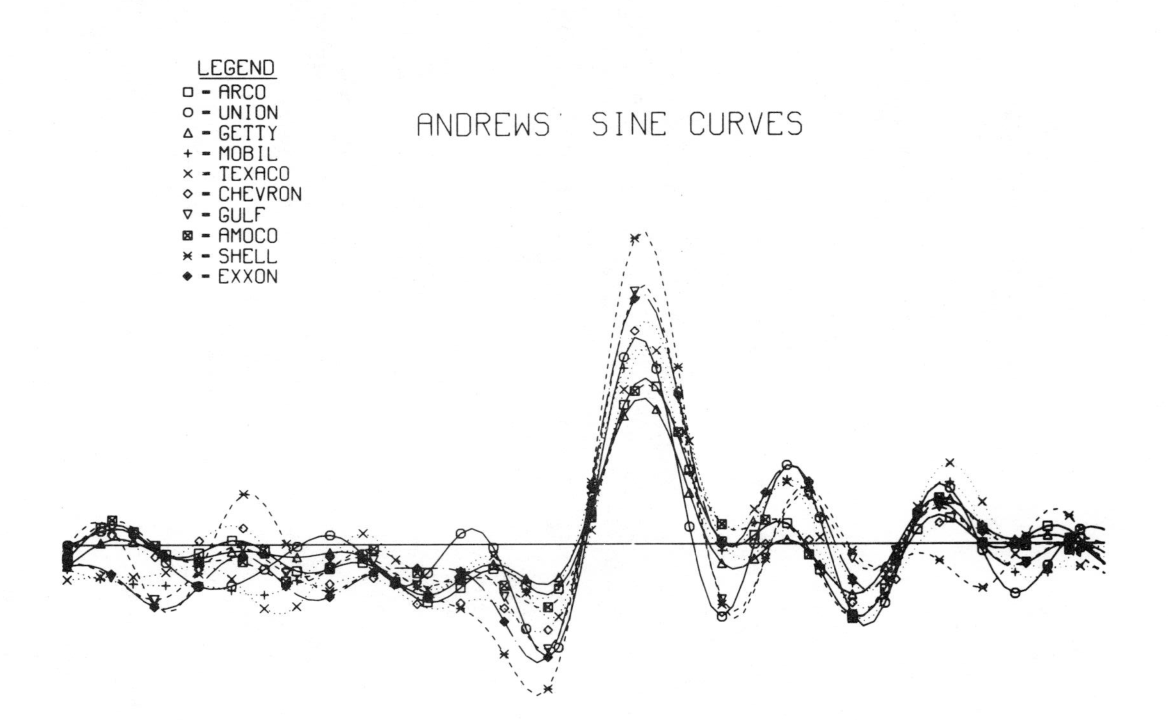

Figure 5. Andrews' sine curves for 10 major oil company groups

LEGEND

AR	ARCO	CH	CHEVRON
UN	UNION	GU	GULF
GE	GETTY	AM	AMOCO
MO	MOBIL	SH	SHELL
TE	TEXACO	EX	EXXON

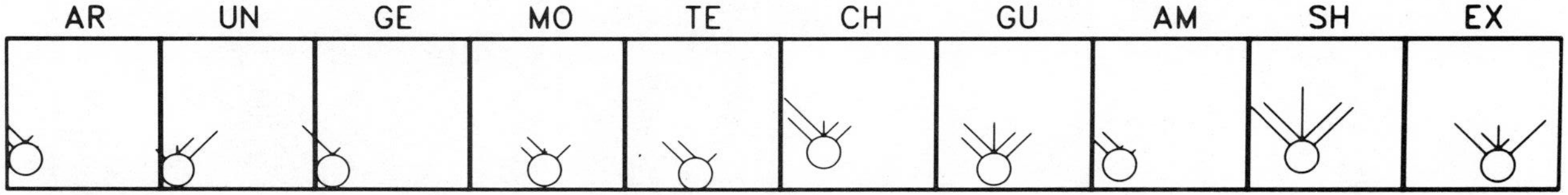

AR UN GE MO TE CH GU AM SH EX

Figure 6. Metroglyphs for 10 major oil company groups: First seven variables

TABLE II. Data on 15 Variables for 10 Oil Company Groups

Company Group Name	Variable														
	1	2	3	4	5	6	7	8	9	10	11	12	13	14	15
ARCO	.56	1.1	.78	306	49	10	4.5	.38	66	62	.11	174	.84	2.8	35
UNION	.53	1.2	.49	203	47	4	4.2	1.22	103	99	.19	527	.98	8.5	38
GETTY	.54	1.0	.32	197	31	11	4.0	.67	51	57	.11	160	.38	2.8	26
MOBIL	1.21	2.8	.50	211	50	8	3.9	1.04	68	78	.06	339	.81	4.0	25
TEXACO	1.16	2.7	.56	176	66	8	7.8	.31	56	50	.04	277	.91	2.5	34
CHEVRON	.84	1.2	1.16	378	70	13	5.8	.70	197	141	.17	355	.50	1.6	32
GULF	1.01	2.2	.67	219	65	11	4.1	1.53	338	235	.23	481	.83	2.9	37
AMOCO	.66	1.3	.66	258	53	8	7.3	.45	37	44	.07	213	.31	2.7	30
SHELL	.97	1.7	1.59	336	95	13	3.6	1.90	430	378	.39	656	.38	2.7	54
EXXON	1.44	2.9	1.02	250	84	8	5.2	.99	276	199	.14	609	.58	4.3	36

particular code for metroglyphs was written by Herbert T. Davis, Jr., of Sandia Laboratories, Albuquerque, New Mexico.

The Andrews' sine curves, the metroglyphs and the Chernoff faces are three different ways of graphically representing multidimensional data. Some of the advantages and disadvantages of each will be discussed in the next section.

IV. ADVANTAGES AND DISADVANTAGES OF USING CHERNOFF FACES

Each of the techniques used to display multidimensional data has advantages and disadvantages associated with its use. The Chernoff face method has several distinct advantages over other representational techniques such as those presented in the previous section.

First, faces are easily recognized and described. We grow up studying faces and learning to recognize different facial expressions. Professor Chernoff has indicated to me that he chose faces over, say houses, because of our ability to differentiate among the former. Differentiation among metroglyphs or Andrews' sine curves is more difficult. It is not even clear how to describe similarity of sine curves.

When a face is presented we can rely on a commonality of language in our discussions. We speak of nose length or ear height and there is no confusion. Metroglyphs can be described, but not quite as easily. Both of these have the advantage of linking individual data variables with figure characteristics. However, this linkage may not always be meaningful.

This leads to a second advantage of using Chernoff faces. We are able to link facial characteristics with the physical meaning of the variables. The smile can be used to represent a "success/failure" variable, the eyes can represent a "slyness" variable or a political stance, the forehead may represent intelligence as was done by Lt. Gerald Lake in a study here at the Naval Postgraduate School. Research on the perception of facial features is shedding light on appropriate uses.

Unfortunately this may make the use and interpretation of faces more subjective. But is subjectivity entirely bad? I do not think so and list subjectivity as the third advantage of using Chernoff faces. The subjectivity is obvious and this distinguishes the face methodology from other techniques. If we are using the faces for clustering, the clusters formed will be influenced by the facial feature-data variable assignment and by the biases of the viewer. If we use a computer package, the choice of clustering algorithm is a subjective choice. Unfortunately, in the latter case it is all too easy to think that the results are arrived at objectively. This is not likely to happen with the face usage. The metroglyph type representations appear objective, but I'm not sure yet. Do we know that we will get the same clustering no matter how the figures are rotated? The Andrews' sine curves will vary with different orderings of the input data.

It must be remembered that there is no universally accepted correct and true method to arrive at clusters. The faces are not being proposed as a method of arriving at final decisions, but rather as a means of studying the data. If the subjectivity of the faces causes the user to be more careful in his or her conclusions, that is fine. If the apparent objectivity of a technique caused the user to treat the technique as final, that would be a disadvantage.

The fourth and final advantage I will give for using faces is that it is possible to concentrate on subsets of the data variables without redoing the graphics. We might want to concentrate on the variables associated with the eyes and ears and then concentrate on the variables associated with the ears and mouth. This concentration is virtually impossible if one uses Andrews' sine curves, and not as convenient when using metroglyphs. The latter suffers from a description problem.

In spite of the above pluses for using Chernoff faces, there are some minuses. Perhaps the first disadvantage to using the faces is that a plotting device is required if one is to draw a standard Chernoff face. I say standard because of a paper (4) by Turner and Tidmore presented at the 1977 Chicago meeting of the American Statistical Association. They demonstrated how Chernoff-type faces can be drawn with a line printer.

A second disadvantage is that a new chapter on the use of Chernoff faces to deceive could be added to Darrell Huff's How to Lie With Statistics. (5) The faces can be abused. However, if we refused to use any technique which can be misused, there would be little left.

A more serious problem with the Chernoff faces is that the built-in dependencies among facial features may distort the data representation enough to cause erroneous impressions. In Section 5, I will discuss this topic more completely. It should be noted that even if all the facial features are independent, there is no guarantee, in fact it is rather unlikely, that the total face will be viewed as a union of 20 different, independent variables.

The final point I'd like to make is that as the number of entities to be represented increases, severe difficulties may occur in actually viewing the faces. This will be particularly true if the faces are similar. If there are two or three very different classes, there won't be much difficulty, if any. This same problem will occur with metroglyphs, Andrews' sine curves and most other representational modes when the number of entities is large.

A similar problem occurs if we try to use all 20 dimensions in the Chernoff faces. Viewing becomes difficult. Fifteen variables is a good maximum. Metroglyph type representations also suffer if the number of dimensions gets too large.

V. FACIAL FEATURE DEPENDENCIES

There is a potentially serious problem involved in using the faces to represent data. While some of the facial features depend only on the input data for the corresponding data variable, other facial features are interrelated to some extent. Face height, the three facial eccentricities, eye slant, ear level and ear size are in the former class; most of the remaining features are in the latter. Pupil position does depend on other facial features only to guarantee that the pupil remains in the eye. The mouth structure, however, depends on face height and width, ear level, lower face eccentricity and nose length, as well as on the three mouth parameters. Eye height depends on nose length and face height; eye separation on the upper face eccentricity and face height. These dependencies occur in order to insure proper positioning of the facial features.

The results of the dependencies can be deceiving. Figure 7 shows eight faces in which all parameters except ear level, nose length and lower eccentricity remain constant. Table III identifies the cases. Notice the effect of these three facial features on the mouth length and forehead.

In Chernoff's original program, the faces were normalized so that both face height and width were constant. The normalization reduces the dependencies, but does not eliminate them. It does, however, essentially remove the face height and width variables from consideration. The program DRFACE does not contain the normalization.

Restriction of the ranges of the facial features reduces the dependencies somewhat. Not using face height, the eccentricities of the upper and lower ellipses and nose length would help greatly but would also cost in terms of lost variables. Loss of nose length is particularly undesirable. Perhaps one solution to the problem is to identify non-overlapping regions for the features and then restructure

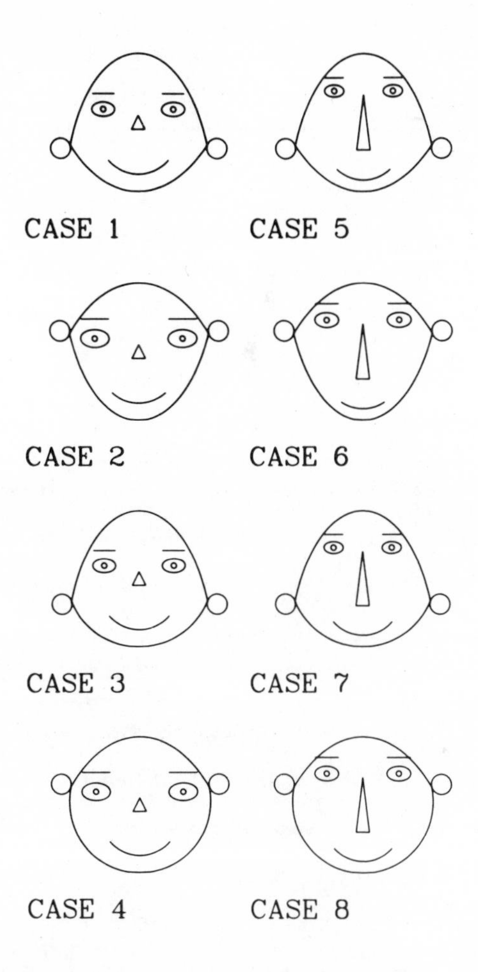

Figure 7. Facial sensitivity to nose length, lower face eccentricity and ear level

TABLE III. Facial Dependency Parameters

Case	Facial Feature		
	Nose Length	Lower Eccentricity	Ear Level
1	.15	.50	.35
2	.15	.50	.65
3	.15	1.00	.35
4	.15	1.00	.65
5	.40	.50	.35
6	.40	.50	.65
7	.40	1.00	.35
8	.40	1.00	.65

DRFACE so that the facial features must lie within these regions. See Fig. 8. Mathematical dependencies would thus be removed. (Perception dependencies may still exist.) There may also be some merit to setting the upper and lower face eccentricities to 1. This will make the face circular.

VI. APPLICATIONS TO CHERNOFF FACES AT LASL

The main application I have made of the faces technique in the past has been to represent data on some of the major oil and gas companies involved in offshore leasing. The ten oil company groups are described in Table IV. (The Arbitrary Company Code (ACC) is a designation given the companies by the Conservation Division, U. S. Geological Survey, Denver, CO.) The variables considered are contained in Table V.

Myrle Johnson of LASL has used the faces program to describe energy related variables on a state by state basis (6). Her paper contains many other examples of the use of graphics to represent data.

Presently I am collaborating with James McFarland and Laird Landon of the University of Houston to use faces to represent quarterly data on nine banks in the Houston area. Faces have been drawn with both random and planned assignment of facial features to data variables. The faces

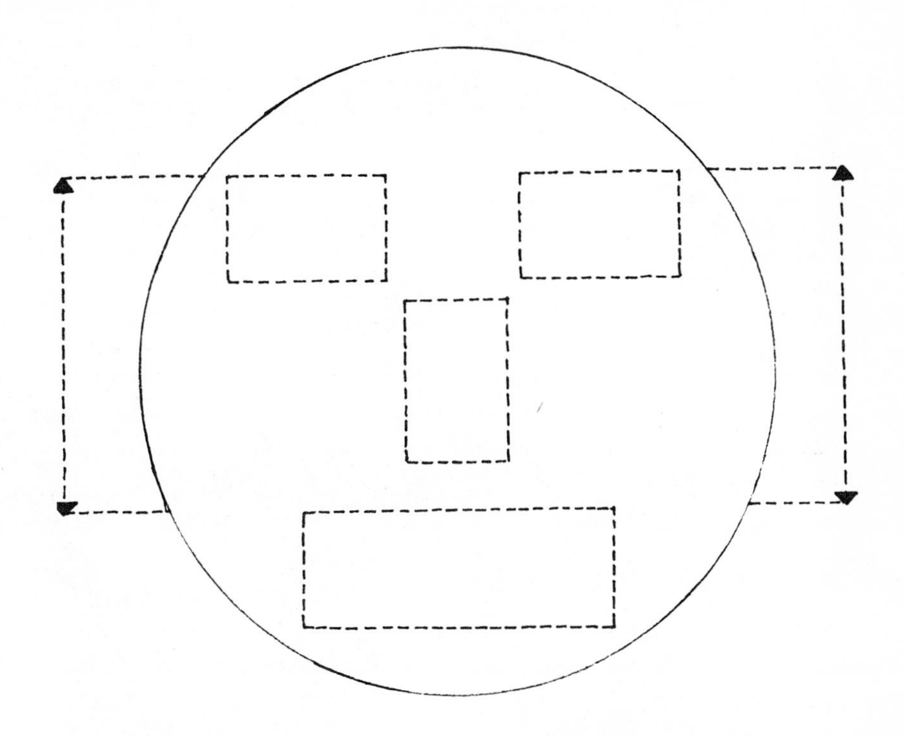

Figure 8. Suggested regions to remove dependencies

TABLE IV. Arbitrary Company Codes (ACC)

ACC	NAME	COMPANIES
2	ARCO	AtlanticRichfield, Richfield Oil, Sinclair, B. B. Barber, Barber Oil Exploration, Royal Gorge Company
3	UNION	Union, Pure Oil, Pure Transportation Company
5	GETTY	Getty, Skelly
39	MOBIL	Mobil, Magnolia Petroleum
40	TEXACO	Texaco, Texaco Seaboard
78	CHEVRON	Chevron, California Company, Standard Oil of Texas
112	GULF	Gulf, British Americal Oil
114	AMOCO	Amoco, Midwest Oil, Standolind, Pan American
117	SHELL	Shell, Shell P/L Corporation
276	EXXON	Exxon, Humble, Exxon Pipeline Company

are to be presented to classes at the University for clustering by students.

Plans are being made to use faces to represent the changing chemical content of water from 17 wells in Los Alamos County. Also it is possible that faces may be used to display the results of an employee attitude survey conducted at LASL the week of February 6-10, 1978. It is proposed to draw one or two faces for each of the Laboratory's 18 divisions.

I would like to discuss briefly one new possible application of Chernoff faces. All the applications discussed so far use faces to display attribute data of some population of interest. I am attempting to see if it is possible to use faces in distributional studies.

The problem being considered is this. Suppose we have a random sample of size 15 from either a normal distribution, N(0,1), or a rectangular distribution on $(-\sqrt{3},\sqrt{3})$. (Both of these distributions have

TABLE V. Description of External Variables

1	Net Bonus	Total net dollars paid (in billions)
2	Excess $/Lease	Average gross dollars paid above 2nd highest bid (in millions).
3	Net acreage	Total net acres leased (in millions)
4	No. leases won	Number of leases won
5	Avg. Ownership	Average percent of ownership of leases
6	Pct. Prod. Leases won	Percentage of leases, ultimately found to be producing, won by the company
7	Avg. Yrs. to Prod.	Average number of years between sale and first production (production lag)
8	Net Gas Prod.	Net gas production (in trillion cubic feet)
9	Net Liq. Prod.	Net liquid production (in millions of barrels)
10	Net Royalty	Net royalty paid to government (in millions)
11	Royalty/Bonus	Net royalty/number of years of production (in thousands)
13	R**2:Roy/Pyr.	Square of correlation coefficient from multiple linear regression of royalty/ prod. yr. on bidding data, production lag and years of production (for producing leases only)
14	Roy/Pyr/Pr.Ac.	Royalty/production year/producing acre (in dollars)
15	Pct. Leases Term-ted	Percentage of owned leases terminated

zero mean and unit variance.) We would like to determine the distribution from which the sample came.

Two approaches are being investigated. In the first, the sample is ordered and the order statistics obtained. Each of the 15 order statistics is assigned to a facial feature, and a face is drawn. This face is then compared to the nominal faces drawn using the same facial feature assignment and the expected values of the order statistics for the normal and rectangular distributions. The distribution whose nominal face is most similar to the sample face is chosen as the parent distribution.

The second approach is to use the face to summarize sample and test statistics and then to compare the sample face to the population faces. Some statistics to be considered are skewness and kurtosis other sample moments, values of the χ^2 and Kolmogorov-Smirnov statistics and perhaps some order statistics or functions of such.

VII. CONCLUSIONS

The Chernoff face technique is one of several available graphical techniques used to display and analyze multidimensional data. When properly employed, it provides useful insight into the nature of the data and has some important advantages over the other graphical techniques. The main criticism of the technique, its subjectivity, can actually be considered as a positive feature rather than as a drawback. The problem of facial feature dependencies can be overcome. In recent years there have been many interesting applications of Chernoff faces in conjunction with cluster analysis, outlier detection methods, distributional studies, and time-series analysis. The Chernoff faces technique will continue to be an effective tool in exploratory multivariate data analysis.

REFERENCES

1. Chernoff, H. (1971). "The Use of Faces to Represent Points in n-Dimensional Space Graphically," Technical Report No. 71, Department of Statistics, Stanford University, Stanford, CA.
2. Andrews, D. F. (1972). "Plots of High Dimensional Data," Biometrics, 28, 125-136.
3. Gnanadesikan, R. (1977). "Methods for Statistical Data Analysis of Multivariate Observations," John Wiley & Sons, NY
4. Turner, D. W., Tidmore, F. E. (1977). "Clustering with Chernoff-Type Faces," Proceedings of the Statistics Section, American Statistical Association Annual Meeting, Chicago, IL.
5. Huff, D. (1954) "How to Lie with Statistics," W. W. Norton & Co.,
6. Lohrding, R. K., Johnson, M. M., Whiteman, D. E. (1978). "Computer Graphics For Extracting Information from Data," to be presented at the 11th Annual Symposium on the Interface. Computer Science and Statistics, March 6-7, Raleigh, N.C.

APPENDIX

```
$FUN(C=S)
$OPEN(FS=IMAGE,SCT=10000)
$OPEN(FS=FILM ,SCT=10000)
$LDGO.
$AFSREL(FS=FSET12,ADISP=TAPE6,POSDEN=556,POSMT=LA350L00)
$FM.
      PROGRAM  DRFACE(INP,OUT,      FSET5=INP,FSET6=OUT,FSET12)
      DIMENSION XFACE(1000),YFACE(1000),YSAME(201),XNOSE(51),
     1YNOSE(51),XMOUTH(51),YMOUTH(51),XLEYE(80),XREYE(80),YEYES(8
     2XLBROW(41),PUPILX(2),PUPILY(2),XRBROW(41),YBROWS(41),
     3Y(40),RANGEY(39),AI(39),BI(39),BIMAI(39),
     4DATA(39),AMODES(200),RAN(2,2), TEMP(39),
     5FEAT(20,2)
      REAL LB, LHSRHS(400),MINY(39),MAXY(39),ID1(1000),L,LSQ
      LOGICAL  IFMT(18),OFMT1(18),OFMT2(18)
      DIMENSION IFOX(39),IRAND(39),IVAR(2)
       DATA  PI/3.141593/,BLANK/1H /
      DATA IFOX/39*0/,NVAR/20/
      DATA Y/40*0.0/,IRAND/39*1/
C     DATA FEAT/.2,.35,.5,.5,.5,.15,.2,-4.,.3,0.,
      DATA FEAT/.57,.35,.5,.5,.5,.15,.2,-4.,.3,0.,
     1.3,.2,.4,.2,.2,.6,0.,.3,.1,.1,
C    2.7,.65,1.,1.,1.,.4,.8,4.,1.,.3,.8,.6,.8,1.,.8,1.,1.,1.,1.,.
     2 68,.65,1.,1.,1.,.4,.8,4.,1.,.3,.8,.6,.8,1.,.8,1.,1.,1.,1.,
      D TA D1,D2/1.,1./
      DATA DATA/0.6, 0.5, 0.5, 0.5, 1.0,0.25, 0.5, 0.0, 0.5, 0.1,
     1          0.7,0.5,0.6,0.5,0.5,0.8,0.5,0.5,0.5,0.1/
      D TA CAPH/1.0/,XINIT/0.0/,YINIT/0.0/
      INTEGER   NU,NL,NFACE,NNOSE,NMOUTH,NEYES,NBROWS
     *,ISTR,IDENT(100)
      DATA NU,NL,NNOSE,NMOUTH,NEYES,NBROWS/400,400,51,51,80,41/
      READ 1,NPLOTS,NFIXED
    1 FORMAT(19I4)
C
C
      WRITE(6,400)
C  READ AND PRINT CONTROL CARDS.
 400   FORMAT( 35X,*CONTROL CARDS READ*//
     $*     FACIAL FEATURE*,25X,*     EXTERNAL VARIABLE*//
     L* NO.  NAME                    LOW     HIGH       NO.  NAME
     $     LOW     HIGH*/ )
      DO 16 IFEAT=1,NFIXED
      READ 410,IVAR(1),TP1,TP2,IVAR(2),TP3,TP4,(RAN(2,J),J=1,2)
  4 0 FORMAT(I4,2A10/40X,I4,2A10,2F8.2)
      JJ=IVAR(1)
      RAN(1,1)=FEAT(JJ,1)
      RAN(1,2)=FEAT(JJ,2)
      WRITE(6,411)(IVAR(1),TP1,TP2,(RAN(1,J),J=1,2),IVAR(2),TP3,T
     1,RAN(2,J),J=1,2))
  411 FORMAT(I4,2A10,2F8.2,5X,I4,2A10,2F8.2)
      KK=IVAR(2)
      IFOX(KK)=IVAR(1)
      KP=IVAR(1)
      BI(KP)=(RAN(1,2)-RAN(1,1))/(RAN(2,2) - RAN(2,1))
    1 AI(KP)=RAN(1,1) - RAN(2,1)*BI(KP)
C
      READ(5,405)IFMT
C     READ IN FORMAT FOR DATA     AND OUT FORMAT FOR DATA
      READ(5,405)OFMT1,OFMT2
  405 FORMAT(18A4)
C
      IPLOT=0
      CALL PLOTS(12)
      CALL PLTZ(0.,-12.,-3)
      YYY=YINIT +1.25
       XXX=XINIT
      CALL PLTZ(0.,1.25,-3)
      NYP=4
COMPUTE VERTICAL   ( Y DIRECTION) NUMBER OF PLOTS
      DO 50 JPLOT4=1,NPLOTS,NYP
 446   FORMAT(18F7.1)
      XXX=XXX+3.
      YYY=YYY+7.50
      CALL PLTZ(3.,7.5,-3)
      ITOP=1
      JEND =MIN0(NPLOTS,JPLOT4+NYP-1)
```

```
      DO 49 JPLOT1=JPLOT4,JEND
      READ(5,IFMT)XID1,(Y(I),I=1,NFIXED)
      WRITE(6,1)
      WRITE(6,1)
      WRITE(6,OFMT1)XID1,(Y(I),I=1,NFIXED)
    2 FORMAT(9F8.2)
      DO 15 IVB=1,NFIXED
      IF(IFOX(IVB).EQ.0)GO TO 15
      KKK=IFOX(IVB)
      DATA(KKK)=AI(KKK)+BI(KKK)*Y(IVB)
   15 CONTINUE
      IF(ITOP.EQ.1)GO TO 17
      YYY=YYY-2.50
      CALL PLTZ(0.,-2.5,-3)
   17 ITOP=0
      DO 13 J=1,NFIXED
      JTP=IFOX(J)
   13 TEMP(J)=DATA(JTP)
      WRITE(6,OFMT1)BLANK,(TEMP(J),J=1,NFIXED)
      WRITE(6,OFMT2)BLANK,(IFOX(J),J=1,NFIXED)
C
C
      HSTAR=.5*(1.0+DATA(1))*CAPH
      THSTAR=(2.0*DATA(2)-1.0)*PI*0.25
      SMALLH=.5*(1.0+DATA(3))*CAPH
      X0=HSTAR*COS(THSTAR)
      Y0=HSTAR*SIN(THSTAR)
C
C     COMPUTE  FACE
      CU=.5*(SMALLH+Y0-X0**2/(DATA(4)**2*(SMALLH-Y0)))
      BU=SMALLH-CU
      AU=DATA(4)*BU
      BUSQ=BU**2
      CL=.5*(-SMALLH+Y0-X0**2/(DATA(5)**2*(-SMALLH-Y0)))
      BL=SMALLH+CL
      AL=DATA(5)*BL
      BLSQ=BL**2
      XMAX=X0
      NFACE=NU+NL
C
      NUP1=NU+1
      YSAME(1)=Y0
      LHSRHS(1)=-X0
      NSTEP=NU/2
      NSTPP1=NSTEP+1
      YSAME(NSTPP1)=SMALLH
      LHSRHS(NSTPP1)=0.0
      STPSIZ=(SMALLH-Y0)/NSTEP
      ISTOP=NSTEP-1
C
      DO 5 I=1,ISTOP
      IP1=I+1
      YSAME(IP1)=Y0+I*STPSIZ
      NUMI=NUP1-I
      XPLUS=DATA(4)*SQRT(BUSQ-(YSAME(IP1)-CU)**2)
      IF(XPLUS.GT.XMAX)XMAX=XPLUS
      LHSRHS(IP1)=-XPLUS
      LHSRHS(NUMI)=XPLUS
    5 CONTINUE
C
      XFACE(1)=LHSRHS(1)
      YFACE(1)=YSAME(1)
      NUP2=NU+2
      DO 6 I=2,NSTEP
      XFACE(I)=LHSRHS(I)
      YFACE(I)=YSAME(I)
      IX2=NUP2-I
      XFACE(IX2)=LHSRHS(IX2)
    6 YFACE(IX2)=YSAME(I)
      XFACE(NSTPP1)=LHSRHS(NSTPP1)
      YFACE(NSTPP1)=YSAME(NSTPP1)
C
```

```
      YSAME(1)=Y0
      LHSRHS(1)=X0
      NLP1=NL+1
      YSAME(NSTPP1)=-SMALLH
      LHSRHS(NSTPP1)=0.0
      STPSIZ=(Y0+SMALLH)/NSTEP
      DO 7 I=1,ISTOP
      IP1=I+1
      NLMI=NLP1-I
      YSAME(IP1)=Y0-I*STPSIZ
      XPLUS=DATA(5)*SQRT(BLSQ-(YSAME(I)-CL)**2)
      IF(XPLUS.GT.XMAX)XMAX=XPLUS
      LHSRHS(IP1)= XPLUS
      LHSRHS(NLMI)=-XPLUS
    7 CONTINUE
C
      NLP2=NL+2
      XFACE(NUP1)=LHSRHS(1)
      YFACE(NUP1)=YSAME(1)
      DO 8 I=2,NSTEP
      XFACE(NU+I)=LHSRHS(I)
      YFACE(NU+I)=YSAME(I)
      IX2=NLP2-I
      XFACE(NU+IX2)=LHSRHS(IX2)
      YFACE(NU+IX2)=YSAME(I)
    8 CONTINUE
      XFACE(NU+NSTPP1)=LHSRHS(NSTPP1)
      YFACE(NU+NSTPP1)=YSAME(NSTPP1)
      XMIN=-XMAX
      YMAX=SMALLH
      YMIN=-SMALLH
C
C     COMPUTE  NOSE
      AN=SMALLH*DATA(6)
      XNOSE(1)=0.0
      XNOSE(2)=SMALLH*DATA(20)
      XNOSE(3)=-XNOSE(2)
      YNOSE(1)=AN
      YNOSE(2)=-AN
C
  9       CONTINUE
C     COMPUTE  MOUTH
      YM=-SMALLH*(DATA(6)+(1.0-DATA(6))*DATA(7))
      XOFYM=DATA(5)*SQRT(BLSQ-(YM-CL)**2)
      AX8=SMALLH/ABS(DATA(8))
      AM=DATA(9)*AMIN1(XOFYM,AX8)
      NSTEP=NMOUTH/2
      NMP1=NMOUTH+1
      YMOUTH(NSTEP+1)=YM
      XMOUTH(NSTEP+1)=0.0
      STPSIZ=AM/NSTEP
      X8SQ=(SMALLH/DATA(8))**2
      HBY8=AX8
      IF(DATA(8).LT.0.0)SIGN=-1.0
      IF(DATA(8).GT.0.0)SIGN=1.0
      DO 11 I=1,NSTEP
      XPLUS=-AM+(I-1)*STPSIZ
      XMOUTH(I)=XPLUS
      NMMI=NMP1-I
      XMOUTH(NMMI)=-XPLUS
      YMOUTH(I)=YM+SIGN*(HBY8  -SQRT(X8SQ-XPLUS**2))
   11 YMOUTH(NMMI)=YMOUTH(I)
C
C     COMPUTE  EYES
      YE=SMALLH*(DATA(6)+(1.0-DATA(6))*DATA(10))
       XOFYE=DATA(4)*SQRT(ABS(BUSQ-(YE-CU)**2))
      XE=XOFYE*(1.0+2.0*DATA(11))*0.25
      THETA=(2.0*DATA(12)-1.0)*PI*0.2
      X13=DATA(13)
      L=DATA(14)*AMIN1(XE,XOFYE-XE)
      LSQ=L**2
      SINTH=SIN(THETA)
      COSTH=COS(THETA)
      R=L/SQRT(COSTH**2+SINTH**2/X13**2)
      PUPILX(1)=-XE+R*(2.0*DATA(15)-1.0)
      PUPILX(2)=XE+R*(2.0*DATA(15)-1.0)
      PUPILY(1)=YE+R*(2.0*DATA(15)-1.0)*TAN(THETA)
      RPUP=DATA(13)*DATA(14)/10.
```

```
C
      NSTEP=NEYES/4
      STPSIZ=L/NSTEP
      I1=1
      I2=NSTEP+1
      I3=2*NSTEP+1
      I4=3*NSTEP+1
      U=0.0
      V=X13*L
      XSTAR=-V*SINTH
      YSTAR=V*COSTH
      XX=XE+XSTAR
      YY=YE+YSTAR
      XREYE(I2)=XX
      YEYES(I2)=YY
      XLEYE(I2)=-XX
      XX=XE-XSTAR
      YY=YE-YSTAR
      XREYE(I4)=XX
      XLEYE(I4)=-XX
      YEYES(I4)=YY
      U=L
      XSTAR=U*COSTH
      YSTAR=U*SINTH
      XX=XE+XSTAR
      YY=YE+YSTAR
      XREYE(I3)=XX
      XLEYE(I3)=-XX
      YEYES(I3)=YY
      XX=XE-XSTAR
      YY=YE-YSTAR
      XREYE(I1)=XX
      XLEYE(I1)=-XX
      YEYES(I1)=YY
      I1=I2
      I3=I4
      ISTOP=NSTEP-1
   10    CONTINUE
      DO 12 I=1,ISTOP
      U=I*STPSIZ
      V=X13*SQRT(LSQ-U**2)
      XSTAR=U*COSTH-V*SINTH
      YSTAR=U*SINTH+V*COSTH
      XX=XE+XSTAR
      YY=YE+YSTAR
      I2=I2+1
      I4=I4+1
      XREYE(I2)=XX
      XLEYE(I2)=-XX
      YEYES(I2)=YY
      XX=XE-XSTAR
      YY=YE-YSTAR
      XREYE(I4)=XX
      XLEYE(I4)=-XX
      YEYES(I4)=YY
      XSTAR=U*COSTH+V*SINTH
      YSTAR=U*SINTH-V*COSTH
      I1=I1-1
      I3=I3-1
      XX=XE-XSTAR
      YY=YE-YSTAR
      XREYE(I1)=XX
      XLEYE(I1)=-XX
      YEYES(I1)=YY
      XX=XE+XSTAR
      YY=YE+YSTAR
      XREYE(I3)=XX
      XLEYE(I3)=-XX
      YEYES(I3)=YY
   12 CONTINUE
C
```

```
C       DRAW EYEBROWS
        YB=YE+2.0*(0.3 + DATA(16))   *L*X13
        THSTST=THETA+PI*(2.0*DATA(17)-1.0)*0.2
        COSTH=COS(THSTST)
        SINTH=SIN(THSTST)
        LB=R*(2.0*DATA(18)+1.0)*0.5
        XX=LB*COSTH+XE
        YY=LB*SINTH+YB
        XRBROW(1)=XX
        XLBROW(1) =-XX
        YBROWS(1) =YY
        XX=-LB*COSTH+XE
        YY=-LB*SINTH+YB
        XRBROW(2) =XX
        XLBROW(2) =-XX
        YBROWS(2) =YY
C
COMPUTE EARS
        REAR=(1.0+DATA(19))*SMALLH*.1
        CEAR=HSTAR+REAR
        EARX=CEAR*COS(THSTAR)
        EARY=CEAR*SIN(THSTAR)
C
C       SET PARAMETERS
C
  61       CONTINUE
        CALL SYMBOL(-1.25,-1.25,.2,XID1,0.,10)
C
  62       CONTINUE
C       DRAW FACE
        XV=0.
        YV=0.
        CALL LINE(XFACE,XV,D1,YFACE,YV,D2,NU,1,0,55B,0)
        CALL LINE(XFACE(NUP1),XV,D1,YFACE(NUP1),YV,D2,NL,1,0,55B,0)
C
  63       CONTINUE
C       NOSE
        CALL PLTZ(XNOSE(1),YNOSE(1),3)
        CALL PLTZ(XNOSE(2),YNOSE(2),2)
         CALL PLTZ(XNOSE(3),YNOSE(2),2)
         CALL PLTZ(XNOSE(1),YNOSE(1),2)
C
  64       CONTINUE
C       MOUTH
        CALL LINE(XMOUTH,XV,D1,YMOUTH,YV,D2,NMOUTH,1,0,55B,0)
  65       CONTINUE
C       EYES
        CALL LINE(XLEYE,XV,D1,YEYES,YV,D2,NEYES,1,0,1H )
        CALL LINE(XREYE,XV,D1,YEYES,YV,D2,NEYES,1,0,1H )
C
  66       CONTINUE
C       EYEBROWS
        CALL PLTZ(XRBROW(1),YBROWS(1),3)
        CALL PLTZ(XRBROW(2),YBROWS(2),2)
        CALL PLTZ(XLBROW(1),YBROWS(1),3)
        CALL PLTZ(XLBROW(2),YBROWS(2),2)
C
  67       CONTINUE
C       PUPILS
        CALL CIRCLE(PUPILX(1),PUPILY(1),RPUP,20)
        CALL CIRCLE( PUPILX(2),PUPILY(1),RPUP,20)
C       DRAW EARS
        CALL CIRCLE(EARX,EARY,REAR,20)
        CALL CIRCLE(-EARX,EARY,REAR,20)
   49   CONTINUE
   50   CONTINUE
        WRITE(6,4051) IFMT
 4051   FORMAT(*0INPUT FORMAT IS *,18A4)
        WRITE(6,4052) OFMT1
 4052   FORMAT(*0OUTPUT FORMAT IS *,18A4)
        CALL PLTZ(0.,0.,999)
C
C
C
        STOP
        END
```

```
      SUBROUTINE CIRCLE(XO,YO,RAD,NPTS)
      DELTH=6.283185/FLOAT(NPTS)
      THETA=0.0
      XX=XO+RAD
      YY=YO
      CALL PLTZ(XX,YY,3)
      DO 5 I=1,NPTS
      THETA=THETA+DELTH
      XX=XO+RAD*COS(THETA)
      YY=YO+RAD*SIN(THETA)
      CALL PLTZ(XX,YY,2)
    5 CONTINUE
      RETURN
      END
$FM.
```

```
10  15
 1 FACE WIDTH
18 BROW LENGTH
 3 FACE HEIGHT
11 EYE SEPARATION
15 PUPIL POSITION
 6 NOSE LENGTH
20 NOSE WIDTH
19 EAR DIAMETER
 2 EAR LEVEL
 9 MOUTH LENGTH
12 EYE SLANT
 8 MOUTH CURVATURE
 7 MOUTH LEVEL
10 EYE LEVEL
16 BROW HEIGHT
```

```
 1 NET BONUS          (B$)     .5     1.5
 2 EXCESS $/LEASE     (MM)    1.0     3.0
 3 NET ACREAGE        (MM)     .3     1.6
 4 NO. LEASES WON            175.0  380.0
 5 AVG. OWNERSHIP             30.0  100.0
 6 PCT.PROD.LEASES WON         0.0   15.0
 7 AVG.YRS.TO PROD.            3.0    8.0
 8 NET GAS PRO.      (TCF)     .3     2.0
 9 NET LIQ.PRO.      (MMB)    35.0  430.0
10 NET ROYALTY       (MM$)    40.0  380.0
11 ROYALTY/BONUS       ($)     0.0     .4
12 ROYALTY/PRO.YR    (K$)    170.0  660.0
13 R**2: ROY/PYR=F( )          .3     1.0
14 ROY/PYR/PR.AC.      ($)    1.5     8.5
15 PCT.LEASES TERM"TED        25.0   60.0
```

```
(A7,2F4.0,13F5.0)
(1X,A10,16F7.2)
(1X,A10,16I7)
ARCO     .56 1.1  .73  306   49   10  4.5  .38   66   62  .11  174  .34  2.8   35
UNION    .53 1.2  .49  203   47    4  4.2 1.22  103   99  .19  527  .98  8.5   38
GETTY    .54 1.0  .32  197   31   11  4.0  .67   51   57  .11  160  .38  2.8   26
MOBIL   1.21 2.8  .50  211   50    8  3.9 1.04   68   78  .06  339  .81  4.0   25
TEXACO  1.16 2.7  .56  176   66    8  7.8  .31   56   50  .04  277  .91  2.5   34
CHEVRON  .84 1.2 1.16  378   70   13  5.8  .70  197  141  .17  355  .50  1.6   32
GULF    1.01 2.2  .67  219   65   11  4.1 1.53  338  235  .23  481  .83  2.9   37
AMOCO    .66 1.3  .66  258   53    8  7.3  .45   37   44  .07  213  .31  2.7   30
SHELL    .97 1.7 1.59  336   95   13  3.6 1.90  430  378  .39  656  .38  2.7   54
EXXON   1.44 2.9 1.02  250   84    8  5.2  .99  276  199  .14  609  .58  4.3   36
```

DRFACE OUTPUT

FACIAL FEATURE NO.	NAME	LOW	HIGH	EXTERNAL VARIABLE NO.	NAME	LOW	HIGH
1	FACE WIDTH	.20	.70	1	NET BONUS (B$)	.50	1.50
18	BROW LENGTH	.30	1.00	2	EXCESS $/LEASE (MM)	1.00	3.00
3	FACE HEIGHT	.50	1.00	3	NET ACREAGE (MM)	.30	1.60
11	EYE SEPARATION	.30	.80	4	NO. LEASES WON	175.00	380.00
15	PUPIL POSITION	.20	.80	5	AVG. OWNERSHIP	30.00	100.00
6	NOSE LENGTH	.10	.40	6	PCT.PROD.LEASES WON	0.00	15.00
20	NOSE WIDTH	.10	.20	7	AVG.YRS.TO PROD.	3.00	8.00
19	EAR DIAMETER	.10	1.00	8	NET GAS PRO. (TCF)	.30	2.00
2	EAR LEVEL	.30	.60	9	NET LIQ.PRO. (MMB)	35.00	430.00
9	MOUTH LENGTH	.30	1.00	10	NET ROYALTY (MM$)	40.00	380.00
12	EYE SLANT	.20	.60	11	ROYALTY/BONUS ($)	0.00	.40
8	MOUTH CURVATURE	-4.00	4.00	12	ROYALTY/PRO.YR (K$)	170.00	660.00
7	MOUTH LEVEL	.20	.80	13	R**2 ROY/PYR=F( )	.30	1.00
10	EYE LEVEL	0.00	.30	14	ROY/PYR/PR.AC. ($)	1.50	8.50
16	BROW HEIGHT	.60	1.00	15	PCT.LEASES TERM'TED	25.00	60.00

INPUT FORMAT IS (A7,2F4.0,13F5.0)
OUTPUT FORMAT IS (1X,A10,16F7.2)

ARCO	.56	1.10	.73	306.00	49.00	10.00	4.50	.38	66.00	62.00	.11	174.00	.34	2.80	35.00
	.23	.34	.67	.62	.36	.30	.13	.14	.32	.35	.31	-3.93	.23	.06	.71
	1	18	3	11	15	6	20	19	2	9	12	8	7	10	16
UNION	.53	1.20	.49	203.00	47.00	4.00	4.20	1.22	103.00	99.00	.19	527.00	.98	8.50	38.00
	.21	.37	.57	.37	.35	.18	.12	.59	.35	.42	.39	1.83	.78	.30	.75
	1	18	3	11	15	6	20	19	2	9	12	8	7	10	16
GETTY	.54	1.00	.32	197.00	31.00	11.00	4.00	.67	51.00	57.00	.11	160.00	.38	2.80	26.00
	.22	.30	.51	.35	.21	.32	.12	.30	.31	.33	.31	-4.16	.27	.06	.61
	1	18	3	11	15	6	20	19	2	9	12	8	7	10	16
MOBIL	1.21	2.80	.50	211.00	50.00	8.00	3.90	1.04	68.00	78.00	.06	339.00	.81	4.00	25.00
	.55	.93	.58	.39	.37	.26	.12	.49	.33	.38	.26	-1.24	.64	.11	.60
	1	18	3	11	15	6	20	19	2	9	12	8	7	10	16
TEXACO	1.16	2.70	.56	176.00	66.00	8.00	7.80	.31	56.00	50.00	.04	277.00	.91	2.50	34.00
	.53	.89	.60	.30	.51	.26	.20	.11	.32	.32	.24	-2.25	.72	.04	.70
	1	18	3	11	15	6	20	19	2	9	12	8	7	10	16
CHEVRON	.84	1.20	1.16	378.00	70.00	13.00	5.80	.70	197.00	141.00	.17	355.00	.50	1.60	32.00
	.37	.37	.83	.80	.54	.36	.16	.31	.42	.51	.37	-.98	.37	.00	.68
	1	18	3	11	15	6	20	19	2	9	12	8	7	10	16
GULF	1.01	2.20	.67	219.00	65.00	11.00	4.10	1.53	338.00	235.00	.23	481.00	.83	2.90	37.00
	.45	.72	.64	.41	.50	.32	.12	.75	.53	.70	.43	1.08	.65	.06	.74
	1	18	3	11	15	6	20	19	2	9	12	8	7	10	16
AMOCO	.66	1.30	.66	258.00	53.00	8.00	7.30	.45	37.00	44.00	.07	213.00	.31	2.70	30.00
	.28	.40	.64	.50	.40	.26	.19	.18	.30	.31	.27	-3.30	.21	.05	.66
	1	18	3	11	15	6	20	19	2	9	12	8	7	10	16
SHELL	.97	1.70	1.59	336.00	95.00	13.00	3.60	1.90	430.00	378.00	.39	656.00	.38	2.70	54.00
	.43	.54	1.00	.69	.76	.36	.11	.95	.60	1.00	.59	3.93	.27	.05	.93
	1	18	3	11	15	6	20	19	2	9	12	8	7	10	16
EXXON	1.44	2.90	1.02	250.00	84.00	8.00	5.20	.99	276.00	199.00	.14	609.00	.58	4.30	36.00
	.67	.96	.78	.48	.66	.26	.14	.47	.48	.63	.34	3.17	.44	.12	.73
	1	18	3	11	15	6	20	19	2	9	12	8	7	10	16

A COMPARISON OF GRAPHICAL REPRESENTATIONS OF MULTIDIMENSIONAL PSYCHIATRIC DIAGNOSTIC DATA

Juan E. Mezzich[1]
David R. L. Worthington

Department of Psychiatry and Behavioral Sciences
Stanford University
Stanford, California

An experiment is described in which eleven experienced psychiatrists each fabricated archetypal psychiatric patients of four different diagnostic types, using the 17 variable Brief Psychiatric Rating Scale. The resulting 17 dimensional data vector was plotted for each "patient" using several different techniques. For each technique, 13 judges attempted to sort the representations into the four equal groups that underlie the data. The difficulty experienced by the judges and the success they had in recovering the diagnostic groups is described for each method. Dimensional reduction to representation as a point in the cartesian plane was the best approach. For judges that did well overall, representation of the data vector as a polar graphed function of an angle was the best multidimensional graphical technique, while judges with low overall recovery rates tended to do better with Chernoff's faces. Finally, an experiment combining factor analysis with polar Fourier representation is presented.

I. INTRODUCTION

One of the most interesting new areas of study in psychiatric classification is the analysis of the cohesiveness and differentiability of diagnostic categories, such as mania and paranoid schizophrenia. A frequent study strategy includes, first, the development of archetypal or conceptual patients fabricated by experienced clinicians to exemplify

---

[1]Supported by MCCPC grant 1-HVA-051 from Stanford Medical School and National V. A. grant MIRS 5692.

ISBN 0-12-734750-X

diagnostic categories, and then the analysis of groupings and other structures underlying the data.

Graphical methods for representing multivariate data points may be useful for the detection and analysis of patterns by facilitating their visualization. It is because of this ease of recognition that graphical representation is relevant and promising for the structural study of diagnostic categories.

In the present study we used seven more or less different graphical techniques for representing a set of 44 archetypal patients defined on 17 psychopathological variables and corresponding to four diagnostic categories.

The two basic objectives of this study were:

1. Assessment of the comparative difficulty experienced by human judges in using the graphical methods for grouping the archetypal patients into four groups.

2. Assessment of the actual performance of these judges in using the graphical methods to recover the four diagnostic groups built into the data.

II. METHOD

A. Data Base

The data base used in this study was composed of 44 archetypal psychiatric patients fabricated by 11 experienced psychiatrists. Each psychiatrist fabricated four archetypal patients to represent four diagnostic categories, namely, manic-depressive depressed, manic-depressive manic, simple schizophrenic, and paranoid schizophrenic.

Each archetypal patient was characterized by a 0-6 severity rating on 17 items or variables from the Brief Psychiatric Rating Scale (Overall & Gorham, 1962). The items were somatic concern, anxiety, emotional withdrawal, conceptual disorganization, guilt feelings, tension, mannerisms and posturing, grandiosity, depressive mood, hostility, suspiciousness, hallucinatory behavior, motor retardation, uncooperativeness, unusual thought content, blunted affect, and excitement.

B. Graphical Methods

The seven graphical methods used in the present study are variations on profiles, faces, functions, and the traditional cartesian plot. In particular, we used polygonal linear profiles, polygonal circular profiles, Chernoff's faces, linear Fourier representations, polar Fourier representations, factor scores in two dimensions directly plotted, and a two dimensional solution obtained through ordinal multidimensional scaling.

The *Linear Profile* is perhaps the simplest method of graphical representation of multivariate points. It represents such a point by a polygonal line which connects the various heights corresponding to the values of the variables, which are arranged along a horizontal (or vertical) baseline. In psychiatry and psychology, the linear profile is most commonly exemplified by the MMPI profile.

Figure 1 shows linear profiles representing the 44 archetypal patients. These "patients" are arranged in line according to the diagnostic group to which they belong.

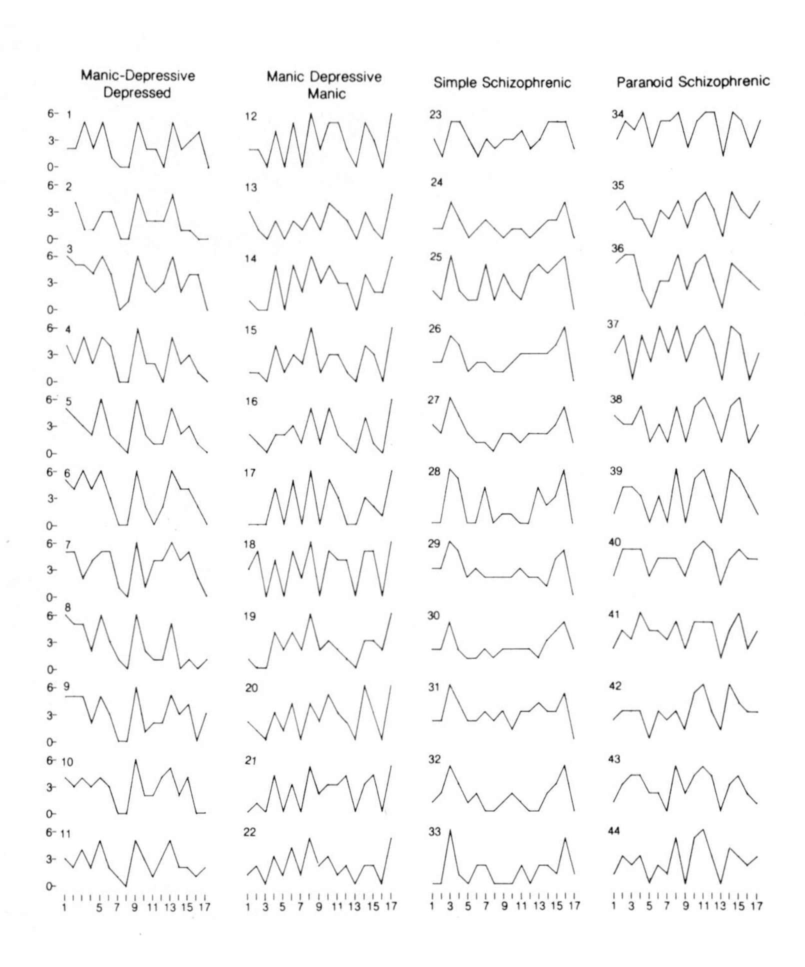

Figure 1. Linear profiles representing 44 archetypal psychiatric patients measured on 17 psychopathological symptoms.

The *Circular Profile* is a variation of the linear profile in which the polygonal line connects points located on equally spaced rays, where the distance from the center represents the value for each of the variables.

Figure 2 shows circular profiles representing the 44 archetypal patients.

Chernoff's Faces (1973) are a very interesting graphical method in which every multivariate point is visualized as a computer drawn face. Each of the features of a face, up to 18 in the version implemented at the Stanford Center for Information Processing, can reflect the value of a variable.

Figure 3 shows Chernoff's faces for the 44 archetypal patients.

In all these methods potentially high dimensional data is put on two dimensional paper in a piecewise manner. Each dimension remains a separate entity and its representation is one of several arranged in a pleasing manner on the page.

Andrews (1972) describes a method that represents high dimensional points as *functions*. Functions of a variable are easily plotted in two dimensions, yet can contain within them arbitrarily high dimensionality. In fact, continuously differentiable functions defined on an interval such as (-3.14,+3.14) define a real vector space, known as $\underline{S}^n$, just as the cartesian plane defines a real vector space, known as $\underline{R}^2$. In part as a consequence of this fact, the functions have several useful properties. The function preserves means, distances, and variances. The graph may show some values of t to be good for discriminating groups. In addition to clustering, tests of significance at particular val-

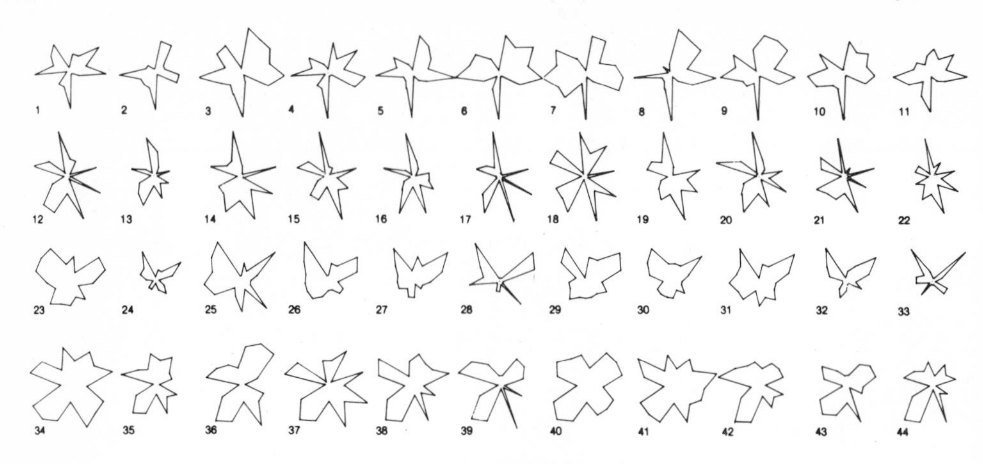

1-11 Manic-Depressive Depressed
12-22 Manic-Depressive Manic
23-33 Simple Schizophrenic
34-44 Paranoid Schizophrenic

Figure 2. Circular profiles representing 44 archetypal psychiatric patients measured on 17 psychopathological symptoms.

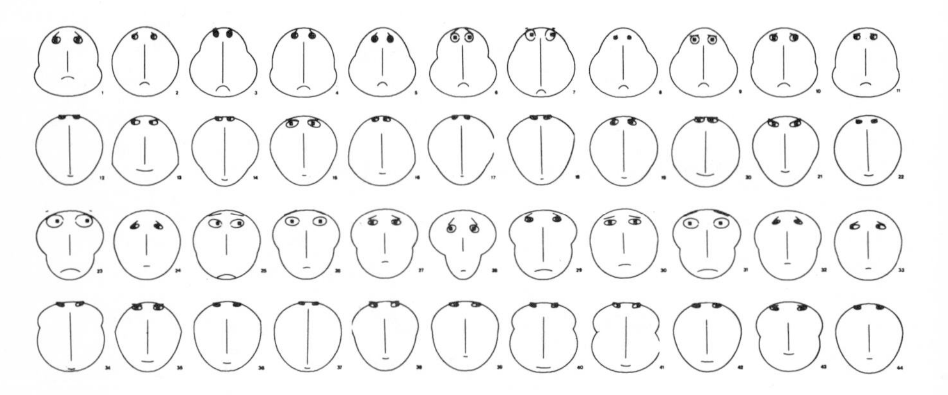

1-11 Manic-Depressive Depressed
12-22 Manic-Depressive Manic
23-33 Simple Schizophrenic
34-44 Paranoid Schizophrenic

Figure 3. Chernoff's faces representing 44 archetypal psychiatric patients measured on 17 psychopathological symptoms.

ues of t can be made, and a confidence region (band) for overall tests can be described. If a point y lies on a line joining x and z then the function it generates will lie between the functions generated by the latter two points.

In this method a Fourier series is used to generate a function of an angle t for each multidimensional point that is to be represented. The function is

$$F(t) = x_1/1.414 + x_2 \cos t + x_3 \sin t + x_4 \cos 2t + \ldots$$

In our case, x_1 to x_{17} are the 17 dimensions that are to be represented.

Figure 4 shows _Linear Fourier_ representations for the archetypal psychiatric patients arranged in columns according to archetypal group. The first term in the equation determines the height of the function, and the others determine its shape. Note should be made of the relative position of the grid lines, which are 0, and the borders. The left and right limits (of t) are -3.1416 and 3.1416, while the top and bottom define the range of the function, which in our case was from -20 to +70. The various functions enter the frame at different values and slopes, and achieve local minimums and maximums at different values of t.

Since it is more difficult to remember horizontal spacing than angular spacing, Chernoff (1973) suggested plotting the functions in polar coordinates. Figure 5 shows the _Polar Fourier_ representations of the 44 archetypal patients.

All of these methods attempt to represent high dimensional data in all of its dimensionality. Another approach to the problem is to reduce the dimensionality of the data to 2, or at most 3, so that $\underline{R}^2$ or $\underline{R}^3$ can be used to represent the data vector exactly as a point.

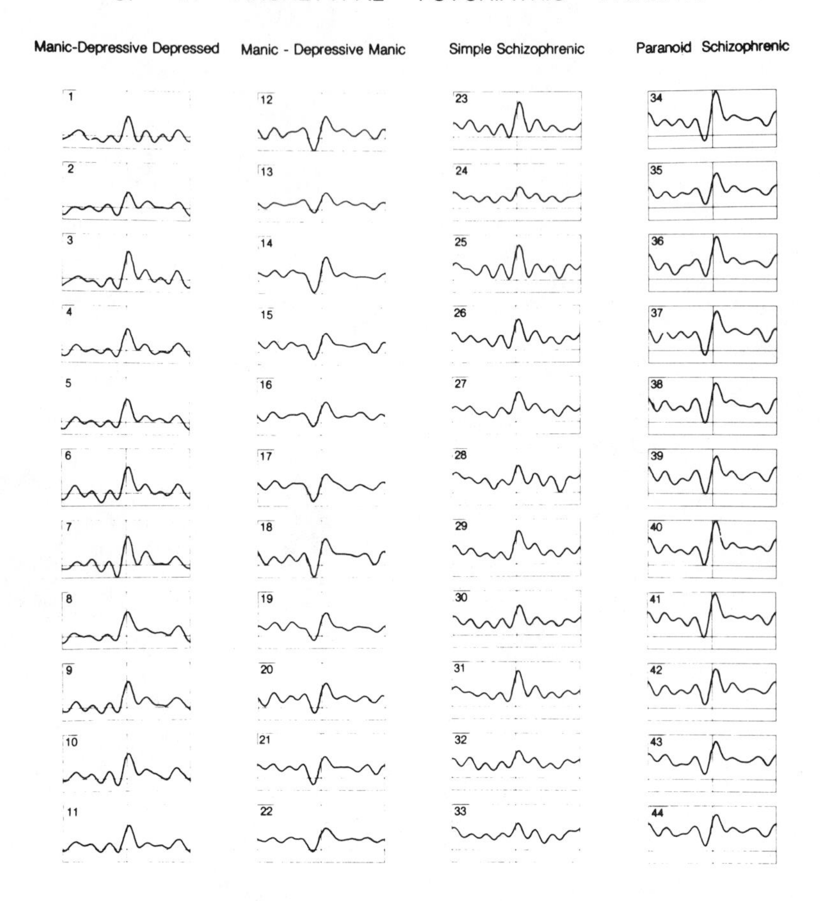

Figure 4. Linear fourier representation of 44 archetypal psychiatric patients.

Factor Analysis is a method which can be used for data reduction. There are many variations of this method. In basic principal component factor analysis the eigenvalues and eigenvectors of the correlation matrix of the original variables are found. Dimensionality is determined from the eigenvalues, and the first few eigenvectors are used as a basis for a lower dimensional representation. Rotation of this solution, orthogonal or oblique, is possible.

Figure 6 shows a plot of the 44 archetypal patients on a space defined by the first two factors obtained by factor analysing the 17 symptoms of our data set. An SPSS (Nie et al., 1970) procedure was used. The default SPSS options of principal component method and varimax rotation were used for this study, but number of factors extracted was specified to be 2. The first factor could be labeled "paranoia" as the variables with highest loadings were suspiciousness, hostility, hallucinatory behavior and uncooperativeness. The second foctor could be labeled "depression" as the variables with highest loadings on it were motor retardation, depressive mood, guilt feelings, and somatic concern. Both factors accounted for 55.6% of the total variance.

Ordinal Multidimensional Scaling (Shepard, 1962; Kruskal, 1964) infers a multidimensional metric structure from nonmetric ordinal data, and represents it in a visualizable, geometrical form, usually on a two dimensional space. The points are arranged in the low dimensional vector space in such a way as to maximize the correspondence of the ranking of inter-point distances in that space to the ranking in the original high dimensional vector space.

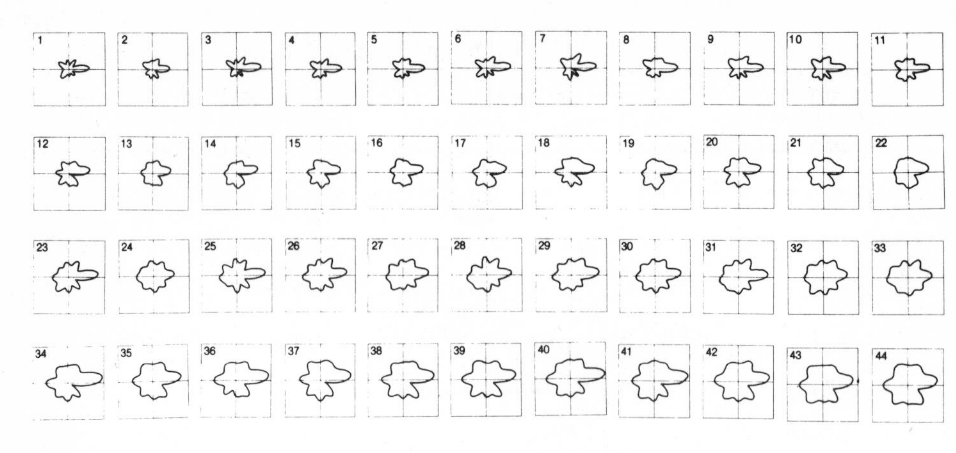

Figure 5. Polar fourier representation of 44 archetypal psychiatric pateints.

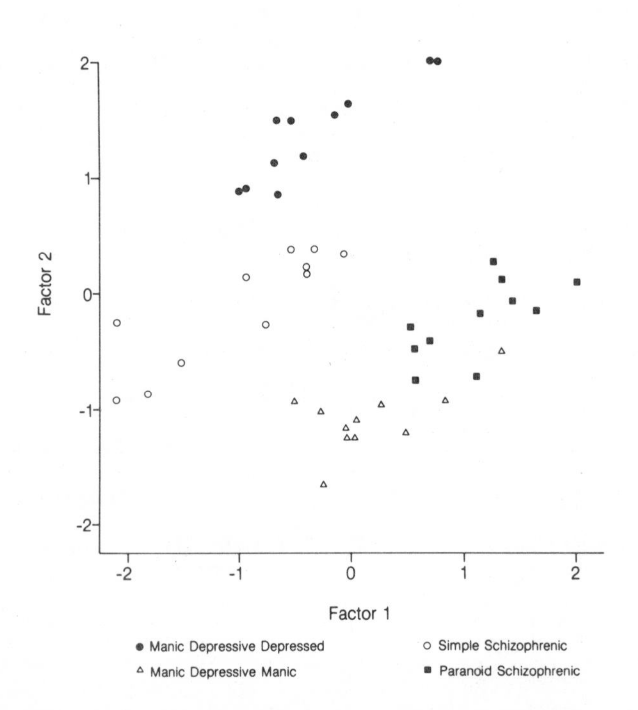

Figure 6. Representation of factor scores for 44 archetypal psychiatric patients on a two factor space.

Figure 7 shows a plot of the 44 archetypal patients on a two dimensional space as obtained by applying ordinal multidimensional scaling, (MDSCAL, version 5M), to our data set. The "stress" level of this configuration was 0.1570.

C. Judges and the Grouping Process

In order to experimentally assess the usefulness of various graphical methods for representing the archetypal patients, 13 people were invited to visually examine the representations shown in the figures, one method at a time, and arrange the 44 representations into four equal groups. All 13 judges were staff members of a psychiatric ambulatory care service, and included three secretaries, three nurses, two social workers, two advanced psychology trainees, one nursing assistant, one psychiatric resident, and an undergraduate student. The packets, containing five envelopes holding 44 (randomly numbered) slips for the first five graphical methods and plain bidimensional plots (with unlabeled points) for the last two methods, were distributed over a three day weekend. As each person returned his packet he was asked to rank the methods by difficulty experienced in deciding on the groupings.

III. RESULTS

Figures 1 thru 7 present the seven types of graphical representation of the 44 archetypal psychiatric patients. Each reader should examine them carefully to appreciate the value of each method. Graphical representations are so recent in comparison with numerical statements that we are just begining to learn how to appreciate the information they contain.

The results of our experimental comparison of the performance of various graphical representation methods are presented in Table I.

TABLE I. Comparative Performance of Seven Graphical Methods

Graphical Method	Ranking of Perceived Difficulty		Percent Agreement with Criterion	
	Mean Rank	S. D.	Mean	S. D.
1. Linear Profiles	5.69	1.44	60.5	11
2. Circular Profiles	5.69	1.03	57.3	15
3. Faces	4.46	1.66	62.9	11
4. Linear Fourier	4.85	1.34	67.8	16
5. Polar Fourier	4.31	1.11	84.1	17
6. Factor Analysis	1.69	0.48	92.0	5.7
7. MDSCAL	1.31	0.48	99.3	1.5

The comparative degree of difficulty experienced by the judges in using the various graphical methods for grouping the archetypal patients is presented in columns 1 and 2 of Table I in terms of mean ranks and standard deviations. These statistics were computed for each method across the ranks provided by the 13 judges. MDSCAL and factor analysis were perceived by all judges as the two easiest or most helpful methods, which is reflected by their mean ranks and standard deviations. The second lot of methods in this regard was headed by polar Fourier representations and ended with both types of profiles.

The second performance aspect comparatively studied was the recovery rate afforded by the various graphical methods, or, in other words, the percentages of agreement with the criterion achieved by the judges using the various methods.

The groupings created by the judges were compared with the intended groupings using a short APL program, which produced cross-classification tables. The recovery percentage for each method and each judge was computed by dividing the trace of the corresponding cross-classification table by 44 and multiplying it by 100. The mean and standard deviation of recovery percentage for each graphical method (computed across judges) are given in the 3rd and 4th columns of Table I. Ten of the 13 judges perfectly recovered the underlying diagnostic groups when using MDSCAL, and this method had the highest mean recovery rate. No instance of perfect agreement with the criterion was obtained with factor analysis representations, but this method had the second highest mean recovery rate. Third came the polar Fourier representations, with which four of the judges perfectly recovered the four diagnostic groups.

Some variability among judges was noted in terms of their mean recovery rates. Mean recovery percentages for the various judges (computed across graphical method) ranged from 64% to 91%, with an overall mean (across judges and graphical method) of 75%. Some inter-judge variability was also noted in ranking patterns. For example, two of the three judges who had the lowest overall recovery rates achieved better recovery with faces than with either the Fourier or profile methods.

The correlation between the mean ranks of perceived difficulty and the mean percentages of agreement with the criterion was -0.93. Again there was variability. All seven judges who thought Fourier easier to use than faces or pro-

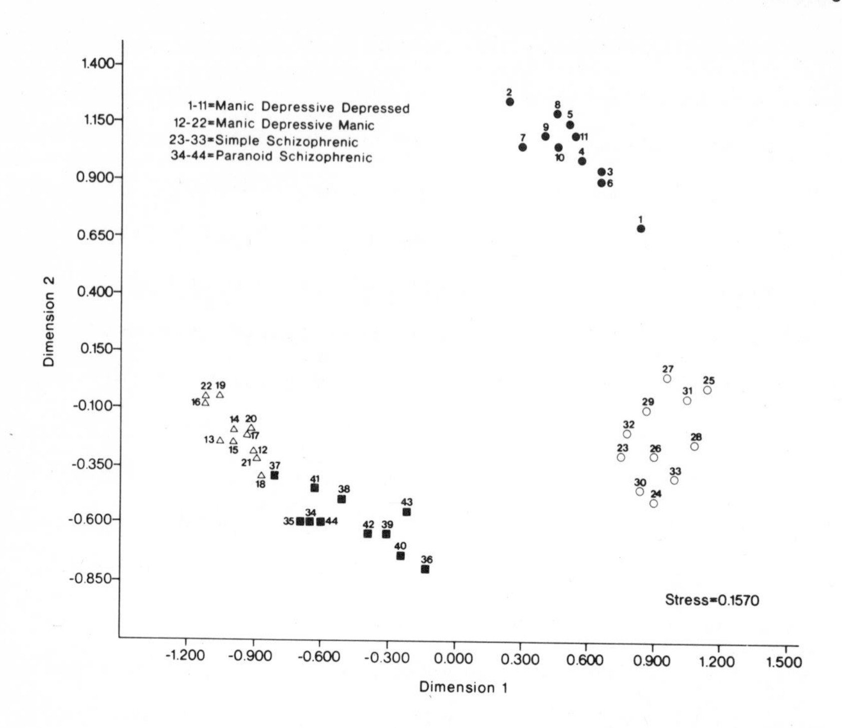

Figure 7. Ordinal multidimensional scaling representation of the 44 archetypal psychiatric patients of data set B.

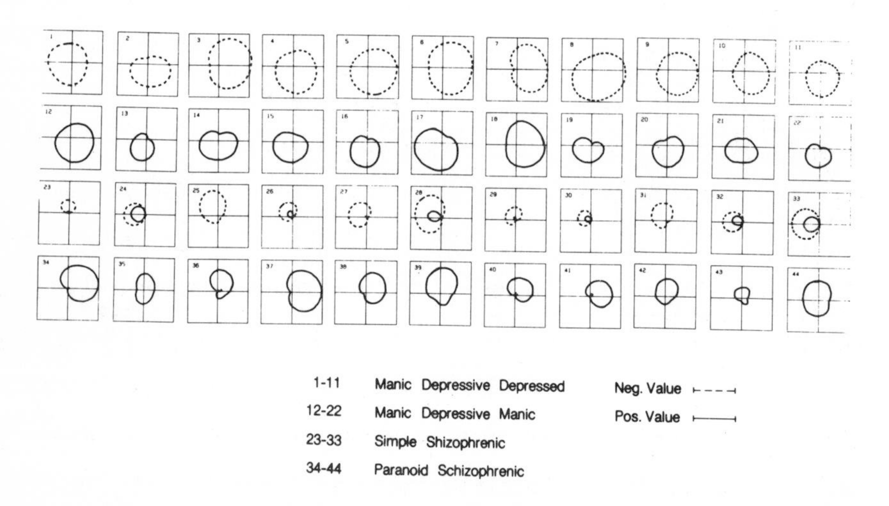

Figure 8. Polar fourier representation of 44 archetypal psychiatric patients using ten factor scores weighted by eigenvalue.

files actually did perform better with Fourier. Two judges who felt faces easier actually did better with that method. However four judges who thought faces easier to use than Fourier actually achieved better recovery with Fourier than with faces.

IV. DISCUSSION

The obvious improvement in recovery of classification found with factor analysis and ordinal multidimensional scaling is probably due in part to the dimensional reduction these methods afford, and in part to our ability to plot the resulting two dimensional points exactly in traditional cartesian coordinates. One of the most interesting findings of this study was the relatively high performance achieved by judges using polar Fourier representations. It seemed reasonable then to combine dimensional reduction through factor analysis with the best performing of the first five graphical methods tested.

Principal component analysis and functional representation are readily combined. In fact the usual problem of assignment of variables to terms in the function (discussed by Chernoff (1973)) is solved by Andrews' (1972) reasonable suggestion of assigning the factor scores in order of importance in accounting for the variance. There are many alternative functions and intervals for producing function plots, but the Fourier series is well known and once that choice was made the selection of interval followed readily. In polar form, negative values of the function must be distinguished, or eliminated as they were in Figure 5 by adding an appropriate constant to the function.

At first the SPSS program described above was run with default program selection of the number of factors. This resulted in a three dimensional solution. The result was plotted with mixed results. With only three terms no second or higher order frequency components entered into the Fourier series and the result in all cases was a more or less circular cardioid. The fine shapes that made recognition at a glance and from a distance possible in the raw 17 dimensional representation exhibited in Figures 4 and 5 were gone here and distinguishing groups became a matter of careful measurement of size and the amount and angle of lean. This can be accomplished by superimposing the curves during their production when the comparisons are known in advance, but for our sorting purposes it was obvious much information was being lost to the casual glance.

It was decided to try a much larger number of factors, obtained through an SPSS image factor analysis program. It produced a ten factor solution. These ten unweighted and unrotated factor scores were plotted as Fourier representations with poor results. The factors extracted late in the process and accounting relatively poorly for the total variance have about the same influence on the Fourier Sum as those more important factors extracted earlier in the factor analysis process. It seems that the higher order terms of the Fourier series, which directly represented the influence of the least important factors, introduced more noise than usable information into the function. This resulted in convoluted forms bearing little resemblance within groups and no discernable differences between groups.

It is appropriate, therefore, when using the Fourier plot with factor scores having roughly the same variance, to weight each score with a value proportional to the amount of variance due to the factor in question. Such weights are easily available in the form of the eigenvalues.

The previous ten factor result was deemed sufficient for a new trial as over 95% of the total variance was accounted for by these factors. The ten factor scores were weighted by their respective eigenvalues and the result is shown in Figure 8. As can be seen, when negative function values are distinguished by a dotted line the groups stand out more sharply than they do in raw form. Widely separated groups, such as the "manic-depressive depressed" and "simple schizophrenic," are clearly and widely separated. Known proximate groups, such as the "manic-depressive manic" and "paranoid schizophrenic," are seen to be more nearly similar. For fine distinctions it is still necessary to superimpose functions, but even without that, known outliers, such as archetypal patients #18 and #37 are easily detected.

The combination of factor analysis and function space plots enables all the dimensionality of a set of data vectors to be utilized with the dimensional information presented in proportion to its importance in accounting for the variance. Graphing factor scores in $\underline{R}^2$ (as in Figure 6) not only drops the higher dimensional information, but with factor scores having roughly the same variance, displays both remaining dimensions equally despite the disparity in their importance as evidenced by the eigenvalues.

Not much research comparing various graphical methods has been reported in the literature. In a previous study mostly comparing numerical clustering methods and using the same data set but a different group of judges than in the present study, Mezzich (1978) found the following recovery rates: 97.7% for MDSCAL, 93.1% for factor analysis (in which a four-Q-factor matrix, rather than a bifactorial plot, was examined), and 59.1% for Chernoff's faces. These figures closely resemble those obtained in the present study for corresponding methods. None of the polygonal profiles or Fourier representations were evaluated in that earlier study.

The variability in performance noted among judges was quite considerable and represents an interesting area for future research. Among the factors most likely to have explanatory power seem to be cognitive style, experience with graphs, and degree of motivation.

ACKNOWLEDGMENTS

We wish to thank B. Balaqui, C. Betz, R. Chalmers, J. Dorenkemper, J. Hejinian, E. Lau, G. Murray, K. Bishop, R. Reed, M. Snodgrass, J. Snyder, D. Theis, and J. Wachtel, who participated as judges in this study; H. Kraemer who reviewed the manuscript; and G. McCue, M. Mills, D. Wonnell and C. Worthington, who helped in various capabilities.

REFERENCES

Andrews, D. F. (1972). "Plots of High Dimensional Data," Biometrics, 28:125-36.

Chernoff, H. (1973). "The Use of Faces to Represent Points in k-Dimensional Space Graphically," Journal of the American Statistical Association, 70:548-554.

Kruskal, J. B. (1964). "Multidimensional Scaling by Optimizing Goodness of Fit to a Nonmetric Hypothesis," Psychometrica, 29:1-27.
Mezzich, J. E. (1978). "Evaluating Clustering Methods for Psychiatric Diagnosis," Biological Psychiatry, 13:265-281.
Nie, N. H. et al. (1970). SPSS: Statistical Package for the Social Sciences, McGraw-Hill, New York.
Overall, J. E. and Gorham, D. R. (1962). "The Brief Psychiatric Rating Scale," Psychological Reports, 10:799-812.
Shepard, R. N. (1962). "The analysis of proximities: Multidimensional Scaling with an unknown distance function," Psychometrika, 27:125-139.

FACIAL REPRESENTATION OF MULTIVARIATE DATA[1]

Robert J. K. Jacob

Naval Research Laboratory
Washington, D. C.

Computer-generated cartoon faces, first proposed by Herman Chernoff in 1971, appear to combine a number of desirable properties for representing multivariate data graphically, including the integrality of the display dimensions and the general familiarity of faces. A series of experiments revealed that, for some useful tasks, the faces do indeed constitute a superior representation for multidimensional Euclidean data. A further series examined how the face displays could be applied to a particular multivariate application. There, it was found that the stereotype meaning already present in faces could be measured and exploited to construct an inherently meaningful display.

I. INTRODUCTION

People are well-known to be proficient at processing visual information (Entwisle and Huggins, 1973). They can do sophisticated processing tasks, almost below the level of consciousness, when the data are presented graphically (Arnheim, 1969). Until the advent of computer graphics, however, people were not nearly as good at generating graphical--or iconic--information as they were at assimilating it. Hence most data

---

[1]This research was supported by a contract between the Johns Hopkins University and the Engineering Psychology Programs, Office of Naval Research; and by the U. S. Public Health Service Hospital in Baltimore, Maryland.

ISBN 0-12-734750-X

were actually communicated using symbols--in the symbolic mode, rather than the iconic mode. Now, the problem has become how best to use the iconic mode for communicating information. While there are some traditional iconic techniques, such as maps and Cartesian graphs, given the new capabilities, it becomes worthwhile to look for new, better, and richer ways to use the iconic mode for communicating information.

A particularly clever iconic device for communicating multidimensional numerical information was proposed by Herman Chernoff (1971; 1973). This was the cartoon human face. Humans look at and process faces constantly. They have become well adapted to this task and are extremely good at performing it. Hence humans would be expected to perform visual processing on faces better than on otherwise comparable visual stimuli. In fact, some evidence suggests that the perception of faces is a special visual process (Yin, 1970).

In order to represent multidimensional numerical data facially, variation in each of the coordinates of the data is represented by variation in one characteristic of some feature of the cartoon face. For example, the first component of the data might be represented by the length of the nose. Other components would be represented by others of the eighteen possible parameters, such as the curvature of the mouth, separation of the eyes, width of the nose, and so on. Then, the overall value of one multidimensional datum would be represented by a single face. Its overall expression--the observer's own synthesis of the various individual features--would constitute a single image depicting the overall position of the point in its multidimensional space. The variety of possible facial expressions would represent the variation possible in a set of numerical data. By looking at the faces and applying one's innate visual processing abilities to them, an observer could perform the visual equivalents of such tasks as multivariate clustering or identifying outliers as easily as he notices family resemblances between people, and by precisely the same, almost unconscious mental mechanism.

Figure 1 shows how the faces are used to represent data. Here, each face represents the value of an eight-dimensional datum chosen from an uncorrelated multivariate normal distribution. One datum differs significantly from the remaining nineteen on several dimensions. It is rather clearly and rapidly identifiable (by a facial expression which differs from the remaining nineteen), despite the presence of considerable noise from the normal distribution. (It is Face 4.)

Several changes were made to Chernoff's original faces for these studies. Most obvious is that the nose was changed from a line to a triangle, and its width is now an additional variable. Chernoff's face height and width parameters were replaced by size and aspect ratio, which better match perceived dimensions. There were some discontinuities and anomalies in the way small changes in the face outline parameters affected the appearance of the face. These were remedied by devising a system of ratio parameters for the face outline. Finally, it was found that reducing the range of variation on most parameters gave a more realistic set of faces; these were preferred because people are especially attuned to very small variations in realistic faces. (The computer program used to generate the faces can be found in Jacob, 1976a or 1976b.)

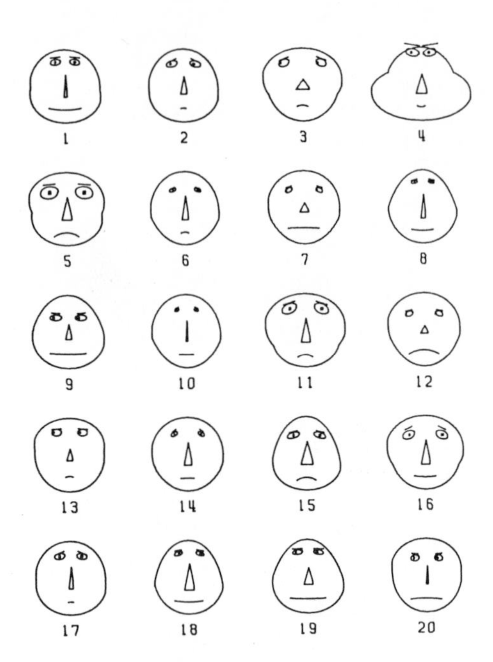

FIGURE 1. Facial display of multivariate data

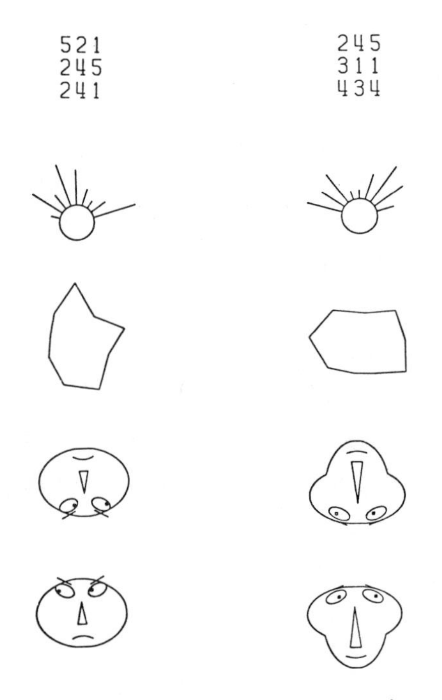

FIGURE 2. Examples from experiment 1

II. PSYCHOLOGICAL EXPERIMENTS

The first set of experiments was intended to ascertain whether subjects could perform common or useful tasks better when the data were displayed as faces or as traditional iconic or symbolic displays. (Jacob, Egeth, and Bevan, 1976 provides more details on these experiments.) In each of the two experiments, subjects were given a simple task to perform involving a set of random data. Performance was compared between subjects who were given the facial representation for the random data and those who were given other representations.

A. Experiment 1

The task in the first experiment was paired-associate learning, a simple, standard task in psychological studies. It consists of asking subjects to learn to associate a name with each data point. Twelve such points were represented by digits, "glyphs" (see Anderson, 1960), polygons (see Siegel, Goldwyn, and Friedman, 1971), upside-down faces, and faces. Each of these displays is illustrated in Figure 2. The entire procedure was repeated for three different dimensionalities. A total of 120 subjects were used.

Results revealed a variety of effects, some mutually confounding. There was a clear dimensionality effect as expected; subjects performed better on points in higher-dimensional spaces, since they contained a greater amount of memorizable information (Egeth, 1966). Because the digit displays lent themselves to rote rehearsal, they induced rather good performance. The overall result, however, was that faces were at least as good as any of the other displays, and often better.

The most interesting observation is that the conventional faces were substantially better than the upside-down faces. Upside-down faces have all the geometrical characteristics of conventional faces, but they lack the familiarity of faces. They were included to determine whether the face is just a geometrically well-designed display (in which case the upside-down face should be just as good) or whether the face is a unique display; results indicated the latter.

B. Experiment 2

While the paired-associate learning task was a standard research task, it was not the sort of task to which the faces were intended to be applied in practice. The second experiment investigated a realistic and practically useful task. This was clustering, or sorting into categories, or pattern recognition.

The task consisted of a set of 50 points in a nine-dimensional space, which were to be organized into 5 groups. They were generated in 5 clusters, each normally distributed around a center point, named the prototype. The subject's task was to look at the 5 prototypes and then assign each of the 50 deviants to a cluster surrounding one of the 5 prototypes. The correct answers were those which put deviants with the prototypes from which they were generated, and to which they were closest in Euclidean distance. While this was a contrived task in that the questions were derived from the answers, it was outwardly similar to many realistic tasks. In a real task, the subject would have the 5 prototypes in his mind, abstracted from his experience or training. He would look at a new data point and assign it to one of the groups he knew. For example, a doctor would examine the data on a patient and then assign him to a cluster which represents a particular disease.

As in the first experiment, the 55 data points were represented in several different ways, and subjects performed the same task with the different displays: faces, a second set of faces with the range of possible variation reduced to three-fourths that of the first set, polygons as in the first experiment, and digits. Figures 3 through 6 present the prototypes (top row of each figure) and examples of their deviants (succeeding rows) for the four different display types respectively. Polygons were used here because they had been found to be the better of the two alternate graphic displays used in the first experiment, probably because their elements are better integrated (Garner, 1974).

Results consisted of the number of errors subjects made in classifying the 50 points. Table I shows the mean number or errors (chance performance would give 40 errors) they made and the mean time (in minutes) they took in sorting the 50 cards. The two types of faces were found clearly to be superior to both the polygons and the digits at $\underline{p} < 0.001$.

TABLE I. Results of Experiment 2 -- 24 Subjects

	Faces	Faces (3/4 range)	Polygons	Digits
Mean no. wrong	15.33	17.08	27.96	31.88
Standard dev.	5.16	5.76	4.98	7.30
Mean time (mins.)	4.14	4.07	3.69	8.24
Standard dev.	1.63	1.46	1.43	3.22

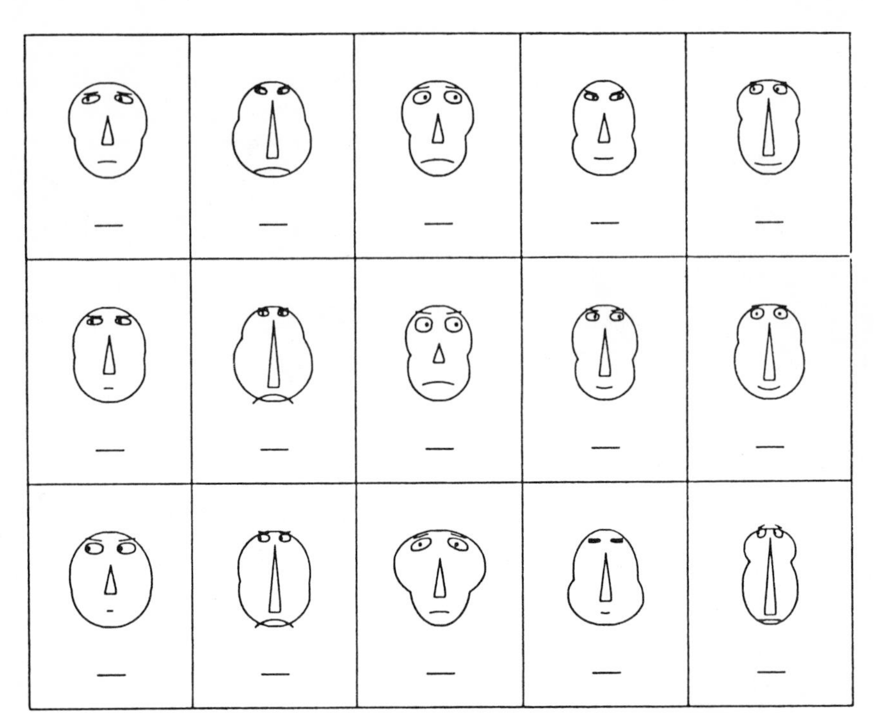

FIGURE 3. Examples from experiment 2: faces

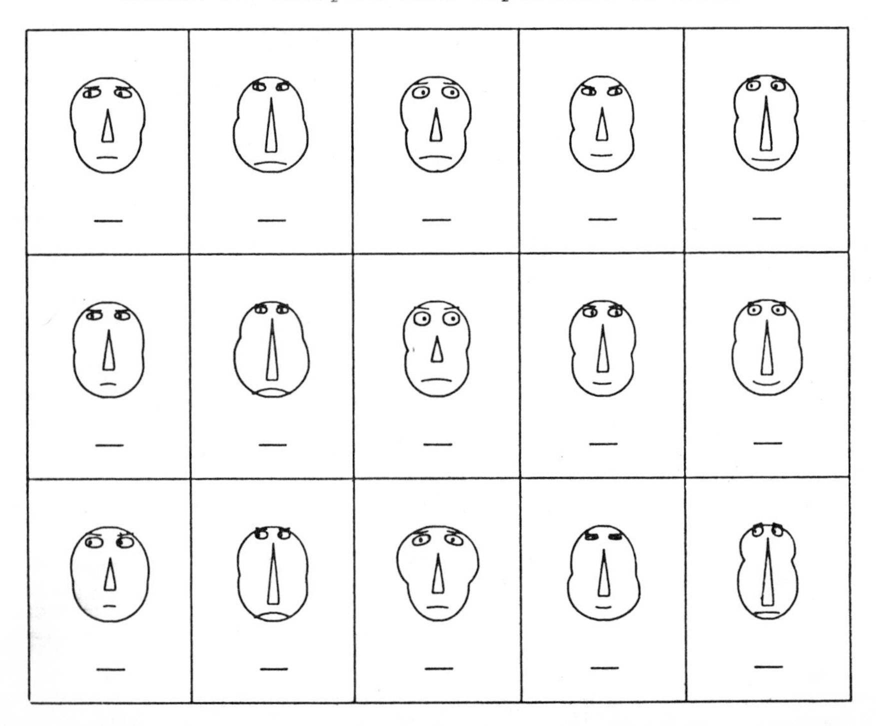

FIGURE 4. Examples from experiment 2: 3/4 range faces

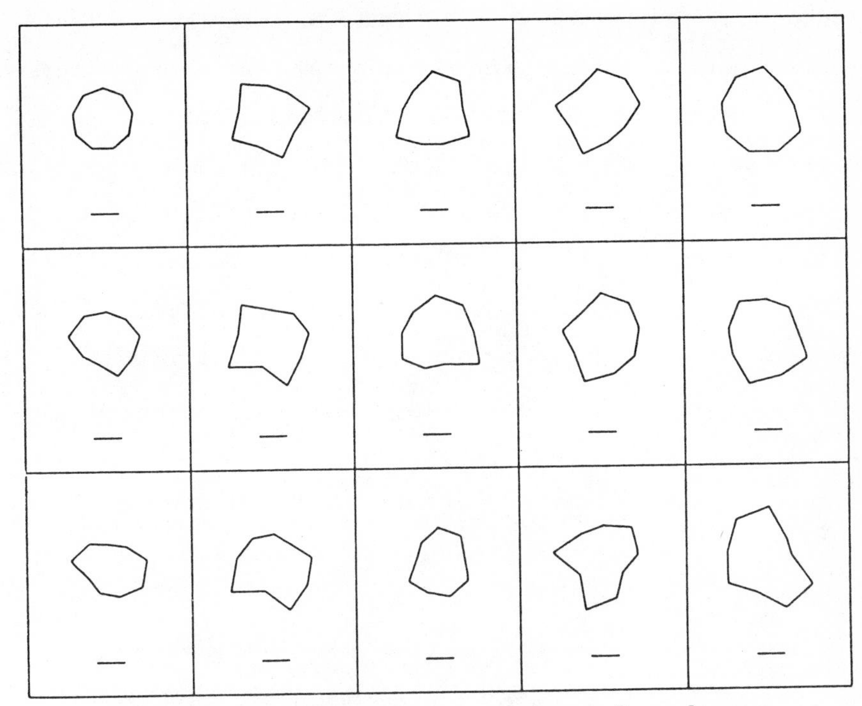

FIGURE 5. Examples from experiment 2: polygons

333 333 333 —	363 636 363 —	633 633 636 —	636 363 366 —	666 666 633 —
345 214 343 —	374 827 454 —	644 532 856 —	635 375 556 —	465 586 734 —
135 123 552 —	345 827 552 —	421 434 424 —	448 285 146 —	886 748 812 —

FIGURE 6. Examples from experiment 2: digits

Significant differences were not found between the two face types or between the polygons and digits. While the polygons could be sorted as quickly as the faces, they were not sorted correctly.

The conclusion drawn was that subjects performed a realistic and useful task significantly better when the data were represented by faces than when they were represented by a conventional display (digits) or by a well-integrated graphic display (polygons). As the experimental task is a fairly general one, one underlying many specific data analysis tasks such as diagnosis, pattern recognition, and cluster analysis, it is claimed that faces provide a superior display for many multivariate applications. Subjects' comments on the experiment help explain this result. They reported that they put all the "happy" faces in one pile, "angry" ones in another, and so on; they found this easy to do. In fact, because of the representation, they were performing a fairly sophisticated multivariate clustering task accurately using only their visual processing abilities. For the other displays, they reported inventing more complicated strategies, which turned out to be self-defeating.

This synthesis by the observer himself of the various graphical elements of the facial display into a single _gestalt_ is one of the principal advantages of this type of iconic display. Many other common types of displays contain several variable elements and could thus be used for graphing multivariate data; but often such displays predispose toward a piecemeal, sequential mode of processing, which obscures the recognition of relationships among elements. By contrast, faces induce their observer to integrate the display elements into a meaningful whole. Previous research with simple cartoon faces and with photographs of real faces has indicated that observers do indeed process these stimuli in such a wholistic fashion (Yin, 1969; Smith and Nielsen, 1970; Reed, 1972).

III. APPLICATION TO MEANINGFUL DATA

Having provided support for the initial supposition that the faces provide a demonstrably good display for Euclidean data of several dimensions, the problem of displaying a specific type of actual (rather than synthetic) data using faces was addressed. The data selected for this purpose were the results of a psychological test intended to determine a patient's psychological personality profile. It was thought that such a profile might possess a more natural facial representation than most other sorts of data. (More details on these experiments, as well as discussions of some related issues, are contained in Jacob, 1976a.)

The form the data took was the results of five particular scales of the Minnesota Multiphasic Personality Inventory (MMPI; Hathaway and McKinley, 1942). The U. S. Public Health Service Hospital in Baltimore administers this test to patients as part of a comprehensive health testing and evaluation and was interested in alternate ways to display the test results. The hospital uses five of the clinical scales of the MMPI: Hypochondriasis, Depression, Paranoia, Schizophrenia, and Hypomania.

Following the approach both of Chernoff and of the previous two experiments, the five components of an MMPI data point could simply have been assigned arbitrarily to five of the facial features (while the unused features were kept at constant values). The resulting facial expressions and the personality traits or disorders which each represents would then be learned by doctors, just as they have learned the meanings of the numerical data and the graphs presently in the patient reports. However, it has been widely observed (e.g., Secord, Dukes, and Bevan, 1954; Harrison, 1964) that particular facial expressions tend to signify particular personality traits to observers with great consistency. Therefore, if the face displays could be devised in such a way that the expression on the cartoon face suggested the same personality traits as those in a particular MMPI report, the resulting face displays would tend to communicate the meaning of the data they represent intuitively. To this extent, a self-explanatory display would have been constructed, somewhat like an hypothetical graph in which it is not necessary to label the axes, because the meaning of the curve is inherently obvious.

Consider, for example, a particularly unfortunate arbitrary assignment of MMPI scales to facial features, in which a smile on the face signified a patient suffering from severe depression. While this could certainly be learned, just as the letters depression or the shape of the personality profile graph are learned, such training would clearly be a poor utilization of the observer's skills.

Therefore experiments were undertaken to attempt to obtain a positive relationship between the 5 components of the MMPI score vector and the 18 variable parameters of the face construction. It was hoped that the resulting face displays would be highly intuitive and suggestive; unlike most computer output formats which require the human observer to learn to understand the computer's language, the power of the computer would here be used to tailor the display format to suit the person's intuition and preconceptions.

A. Experiment 3

The first experiment in this study attempted to measure a relationship between MMPI scores and face parameters based on one observer's preconceptions or stereotypes. This corresponded to a transformation between the 5-space of MMPI scores and the 18-space of face parameters. Because of the imprecision in the process of perception of personality from faces, it was hoped that a linear model would be sufficiently accurate for useful results. Subsequent analysis of the experimental results for higher-order interactions showed this to be a reasonable choice. Moreover, the dimensionality of the problem made any other model very much more difficult to study. Thus a matrix (T) was proposed to define a linear transformation from the space of MMPI score vectors (d for diagnosis) into that of face parameters (p).

A set of 200 faces was generated using parameter (p) vectors chosen from an 18-variate uniform random distribution. Figure 7 shows a sample of these faces. Dr. Faith Gilroy, a research psychologist at the Public Health Service Hospital, then rated each of the faces on the five scales. She was, in effect, indicating what MMPI results each of the faces signified to her, or, more specifically, what MMPI score she thought a person who looked like each of the 200 faces would receive.

A multiple linear regression of the p vectors on the d's was computed from 200 pairs of such vectors, producing a T matrix of regression coefficients (Jacob, 1976a). That matrix could then be used to estimate a p vector (or face drawing) for any given d vector (or personality score). Such estimated p vectors were computed and compared to the original (stimulus) p vectors; the mean squared error over all components of all the vectors was 0.07497 (components of the p vectors ranged between 0 and 1).

The T matrix was displayed graphically by computing the p vectors which correspond to equally-spaced points along the axes of the d space (that is, points which represent patients who have only one psychological disorder). Figure 8 shows the resulting display. In it, each row depicts a series of patients with increasing amounts of a single disorder. Because of the rating scale used, 0 (the first column) represents an inverse amount of the disorder, 1 represents no disorder (the origin of the d space), and 4 represents a large, extrapolated amount of the disorder. It was thought that these faces (particularly those in the column labelled 3) actually corresponded to common stereotypes of the personality traits they were claimed to represent. The subject had never seen these faces nor any resembling them; rather, they had been deduced from the linear regression using faces reported to have more than one disorder.

Some comparisons were made between this T matrix and results obtained by previous investigators. While no studies had used stimuli of this complexity or the same rating scales, some of the observed relations between basic facial feature variations and basic emotions were confirmed. Comparison to the work of McKelvie (1973) and of Harrison (1964) corroborated both the major axes of facial variation found in the T

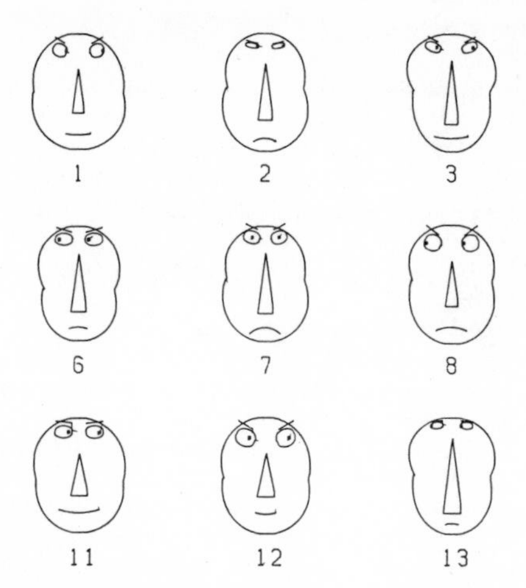

FIGURE 7. Examples from experiment 3

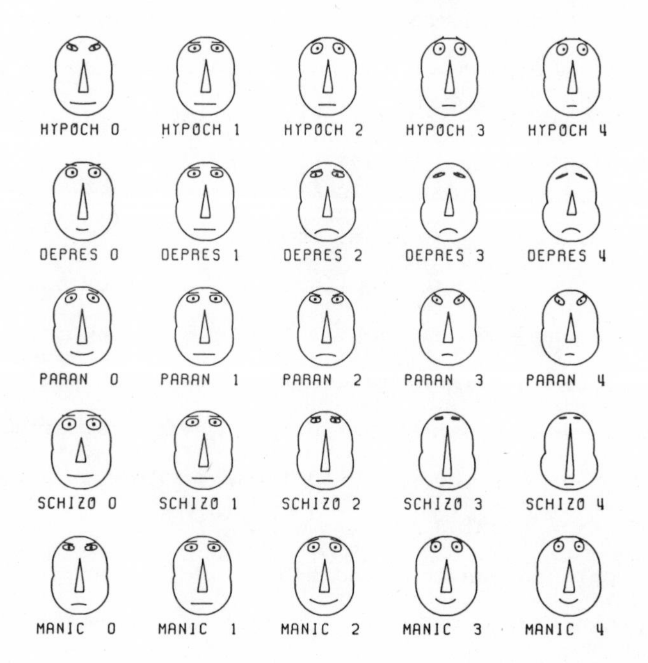

FIGURE 8. Facial representation of the T matrix

matrix and their relationships to variation in emotional states. (As one might expect, these all suggest that the joint variation in the mouth and eyebrows are the major determinants of emotional content of the facial expression; that variation induces variations along axes comparable to Paranoia, Depression, and Hypochondriasis.)

An additional computation showed that the angles in the p space between the facial representations of the orthogonal axes of the d space ranged from 70 to 112 degrees, suggesting that the orthogonal d axes were indeed perceived as being related to orthogonal variations in their facial representations.

Thus, Experiment 3 provided a linear transformation from MMPI scores to faces which was both intuitively appealing and internally consistent. Further study was undertaken in order to validate and then apply this relationship.

B. Experiment 4

An attempt was made to replicate the previous experiment with the same and with another subject. A new set of random faces, generated similarly to the first set, was presented to two subjects who rated them as in the previous experiment. Actual responses were compared to those predicted using the T matrix of Experiment 3.

The comparison was confounded by the appearance of significant response bias. That is, subjects gave consistently higher or consistently lower ratings to the faces on certain scales. It could be determined that, in those cases where the response *magnitudes* matched the predictions (approximately half of the data), the present results supported the previous ones in *direction* as well. In the remaining cases, neither support nor contradiction could be asserted. This experiment could have been improved by embedding the stimulus faces in a larger group which would have induced subjects to attain the same mental set (and thus the same response bias) as that of the subject during Experiment 3. Instead, the insights gained from this experiment were used to devise a new experiment which would provide a more powerful test of the transferability of the T matrix relationship.

C. Experiment 5

For the relationship T to be valid and transferable to other observers, it must appeal to intuitive stereotypes which are already present in the minds of most observers. Such stereotypes need not possess any absolute validity; they need only be widely and uniformly held in order to be exploitable in devising a facial display for MMPI scores. Thus, this experiment was designed to test the applicability of the stereotypes already discovered. Untrained subjects in the experiment were asked to match facial representations of random hypothetical MMPI scores to alternate representations of the same data. Since the numerical MMPI scores were not meaningful to the subjects (or to the intended final users of the display), an independently-developed textual representation for MMPI scores (Rome *et al.*, 1962) was used in this study.

The 30 subjects were each given 50 stimuli, an example of which appears in Figure 9. In each, the subjects were asked to indicate which of the five faces given best corresponded to the given text description. In fact, that description was the textual representation of a particular point in the d space.

NO. 22

-RESENTFUL AND SUSPICIOUS OF OTHERS

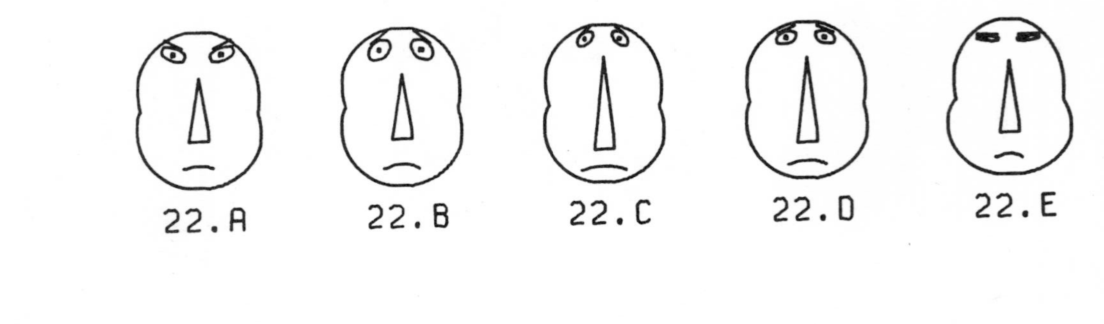

BEST CHOICE-

FIGURE 9. Example from experiment 5

One of the five faces was the facial (using the T matrix) representation of the same point, and the remaining faces were representations of other, randomly-selected points.

The principal result of interest was whether entirely naive subjects could select the face which was claimed (by the results of Experiment 3) to represent the same MMPI data as the text at better than chance performance. If the T matrix had no wider validity than for one subject at one point in time, the subjects would not perform the present task; if, however, the matrix relationship corresponded to widely-held stereotypes, the subjects would use such to perform this task better than a random guessing hypothesis would predict. Results were obtained by measuring the Euclidean distance in the 5-dimensional d space between the expected answer and the answer a subject chose. Such a distance could range from 0 (correct choice) to 4.5 (the maximum diagonal dimension of the hypercube). Table II presents these data. A matched *t* test on the data revealed that subjects were able, with highly significant ($p < 0.0005$) accuracy, to choose those faces which were designed to communicate the same information as the text items.

Such a result suggests that the faces plus the T transformation obtained provide a data display which requires no

TABLE II. Results of Experiment 5 -- 30 Subjects

Chance performance	1.571
Mean observation	1.226
Standard deviation	0.146
t_{29}	12.975

training of the observer. Without any prior information other than their innate facial stereotypes, subjects were able correctly to perceive the data being displayed.

D. Experiment 6

Experiment 5, then, demonstrated that the faces could be used to communicate psychological data to naive subjects. Experiment 2 showed that a particular useful task could be performed better and more quickly with facially-represented data than with several other representations. Together, the experiments suggest that the face might be a superior mode of displaying the MMPI data under consideration. The present experiment was intended to test this composite hypothesis by having subjects perform a meaningful and realistic task which requires apprehension of MMPI data. Various subjects would perform the same task using the facial and the textual representations of the same MMPI data, and their performance would be compared.

A truly realistic task would be the diagnosis and treatment of a real patient; the results would be measured by evaluating the patient's well-being at the conclusion of the treatment. Unfortunately, there would be far too many confounding variables in such an experiment (as well as ethical considerations). Instead, a crude task, analogous to psychological triage, was devised. Subjects were asked to rate the overall emotional well-being of an hypothetical patient, given his MMPI test scores presented in one of two ways. Their success would be measured by comparing their responses to the responses of a clinical psychologist who studied the unprocessed numerical MMPI scores. Thus, to the extent that a naive subject's responses, using the facial or textual representation, corresponded to this baseline, it could be claimed that, through the use of that representation for the data, he was able to perform the same task as the trained psychologist.

Thirty-two subjects were each given fifty stimuli, each of which resembled either Figure 10 or Figure 11. In each case, the subject was being asked to rate a random point in the d space (represented facially or textually) for emotional well-being.

Results were obtained by measuring the correlation coefficient between a subject's ratings and those of the psychologist. A chance hypothesis would have predicted zero correlation. The mean correlation scores over subjects are presented in Table III. First, one can observe that subjects' performance exceeded chance expectation significantly ($p < 0.005$) for both faces and text. Next, conventional and also paired-observation t tests were made to find the difference

TABLE III. Results of Experiment 6 -- 32 Subjects

	Text	Faces
Mean correlation score	0.644	0.399
Standard deviation	0.095	0.111
Difference from chance--t_{31}	38.485	20.416
Difference between means--t_{62}	9.533	
Paired observations difference--t_{31}	8.823	
Score using text algorithm	0.667	

EMOTIONALLY WELL ....EMOTIONALLY DISTURBED

FIGURE 10. Example from experiment 6

```
-ABOVE AVERAGE NUMBER OF PHYSICAL COMPLAINTS, UNDUE CONCERN
 WITH BODILY HEALTH
-TOUCHY, UNDULY SENSITIVE, SUSPICIOUS, INCLINED TO BLAME OTHERS
 FOR OWN DIFFICULTIES

                              NO. 31




        I-----I           I-----I           I-----I
        I     I           I     I           I     I
        I     I           I     I           I     I
        I-----I           I-----I           I-----I

EMOTIONALLY WELL.......................EMOTIONALLY DISTURBED
```

FIGURE 11. Example from experiment 6

between the two display types. Both tests showed that subjects performed the task significantly ($p < 0.005$) better when given the text than when given the faces.

Some insight into this unexpected situation may be gained by studying the text displays in more detail. It appeared that the more disturbed a patient was, the longer his text description was. Hence subjects' responses to the text could have been based on this unexpected iconic content of the text display; they could have been responding to the quantity of text rather than to its meaning. To test this, an algorithm which rated the emotional well-being of a patient based only on the quantity of text in the textual representation of his MMPI score was applied to the experimental stimuli. As shown in the table, the algorithm achieved slightly better performance than the subjects who used the text display. Thus the superior performance of the text displays could be explained by their unintentional iconic content; or, illiterate subjects could have produced the same responses from the text displays as did college students.

The conclusions of this experiment are, then, unclear. While the text displays induced better performance, this turned out to be explainable by an irrelevant property they were found to possess. Nevertheless, the usefulness of faces for inducing good performance in processing Euclidean data was established by Experiment 2; and the ability of the transformation discovered in Experiment 3 to transmit data facially without training was established by Experiment 5. These continue to suggest that an improved version of Experiment 6 would indicate superiority for the facial representation.

Experience in constructing Experiment 6, however, suggests that it would not be a trivial task to devise an unassailable experimental test of this hypothesis.

IV. CONCLUSIONS

Two principal conclusions are drawn from this study. First, computer-produced faces are a particularly good representation for inducing superior performance of useful tasks on multivariate metrical data. Experiments with other iconic and symbolic displays indicate that it is the face display itself, not merely the iconic mode, that accounts for this superiority. Second, the stereotype meaning already present in faces can be utilized in constructing a display. It was possible to measure and then exploit such meaning in order to create a demonstrably self-explanatory display for a particular set of data.

ACKNOWLEDGMENTS

Prof. William Huggins was the author's advisor while the author was a graduate student in Electrical Engineering at the Johns Hopkins University; he provided guidance and insight throughout this work. Profs. Howard Egeth and William Bevan at Johns Hopkins guided the work on the first set of experiments. Drs. Richard Hsieh and Faith Gilroy of the U. S. Public Health Service Hospital in Baltimore provided important assistance for the second set of experiments.

REFERENCES

Anderson, E. (1960). Technometrics 2:387.

Arnheim, R. (1969). "Visual Thinking." Univ. of California Press, Berkeley.

Chernoff, H. (1971). Stanford University Department of Statistics Technical Report No. 71.

Chernoff, H. (1973). American Statistical Assoc. J. 68:361.

Egeth, H. (1966). Perception and Psychophysics. 1:245.

Entwisle, D. and Huggins, W. (1973). Child Development. 44:392.

Garner, W. (1974). "The Processing of Information and Structure." Lawrence Erlbaum, Potomac, Md.

Harrison, R. (1964). "Pictic Analysis." Doctoral dissertation, Michigan State University.

Hathaway, S. and McKinley, H. (1942). "The Minnesota Multiphasic Personality Schedule." Univ. of Minnesota Press, Minneapolis.

Jacob, R. (1976a). "Computer-produced Faces as an Iconic Display for Complex Data." Doctoral dissertation, Johns Hopkins University.

Jacob, R. (1976b). "PLFACE Program." Available from the author upon request.

Jacob, R., Egeth, H., and Bevan, W. (1976). Human Factors. 18:189.

McKelvie, S. (1973). Perception and Psychophysics. 14:343.

Reed, S. (1972). Cognitive Psychol. 3:382.

Rome, H. _et al_. (1962). Proc. Mayo Clinic. 37:61.

Secord, P., Dukes, W., and Bevan, W. (1954). Genetic Psychol. Monographs. 49:231.

Siegel, J., Goldwyn, R., and Friedman, H. (1971). Surgery. 70:232.

Smith, E. and Nielsen, G. (1970). J. Experimental Psychol. 85:397.

Yin, R. (1969). J. Experimental Psychol. 81:141.

Yin, R. (1970). "Face Recognition: A Special Process?" Doctoral dissertation, M.I.T.

MULTIVARIATE DENSITY ESTIMATION BY DISCRETE MAXIMUM PENALIZED LIKELIHOOD METHODS

David W. Scott[1]

Department of Community Medicine
Baylor College of Medicine
Houston, Texas

Richard A. Tapia
James R. Thompson

Department of Mathematical Sciences
Rice University
Houston, Texas

The graphical representation of a multivariate random sample by a probability density function is a useful one. The parametric approach to fitting a density function, such as the multivariate Gaussian, is not robust against incorrect specification of the parametric form. Nonparametric density estimation, on the other hand, provides consistent approximations for a large class of well-behaved "true" sampling densities. The authors have previously developed a univariate nonparametric density estimator that is a continuous piecewise linear function that optimizes a discrete maximum penalized likelihood criterion. This estimator is pointwise consistent almost surely and Monte Carlo studies indicate good small-sample approximation properties. In the present paper, we present two methods of extending the univariate algorithm to higher dimensions. In the first method, the criterion is extended directly to higher dimensions. In the second method, the eigenvectors of the estimated covariance matrix are used to provide a pseudo-independent linear transformation. Then the univariate algorithm is applied to each of the pseudo-independent dimensions and the multivariate estimate is easily obtained. The latter method is illustrated with multimodal data.

[1]This work was supported in part by the National Heart and Blood Vessel Research and Demonstration Center through the National Institutes of Health grant number 17269 and by the U.S. Office of Naval Research and the U.S. Energy Research and Development Administration under grants NRO42-283 and E-(40-1)-5046, respectively.

ISBN 0-12-734750-X

I. INTRODUCTION: GRAPHICAL REPRESENTATION OF DATA

In this paper, we consider the statistical problem of providing compact representation of large data sets while preserving the information contained in the data. Some procedures for this dimensionality reduction are as simple as calculating means and covariances, while other procedures, such as plotting histograms, present the data in graphical form. These examples are both common techniques for probability density estimation. In general, a probability density function (pdf) is the best information-preserving transformation for randomly sampled data. Other advanced statistical techniques for describing and comparing data sets generally require information which is contained in the probability density estimates of the data sets. The graph of a multivariate estimate is one of the most useful graphical representations of randomly sampled data sets.

II. NONPARAMETRIC PROBABILITY DENSITY ESTIMATION

Existing techniques for probability density estimation may be divided into two groups: parametric and nonparametric. In the parametric approach the investigator formulates an experimental model which implies a particular form for the "true" underlying pdf. Thus the primary purpose of collecting data is to provide estimates of the unknown parameters in the pdf formula. By far the most common parametric form assumed for multivariate data is the k-dimensional Gaussian distribution, which contains $k+k(k+1)/2$ unknown parameters in the mean vector and

FIGURE 1. Superposition of the discrete maximum penalized likelihood estimates (solid line) and histogram estimates (dotted line) of plasma lipid concentrations for 320 males with coronary artery disease.

FIGURE 1A.

Plasma cholesterol concentration

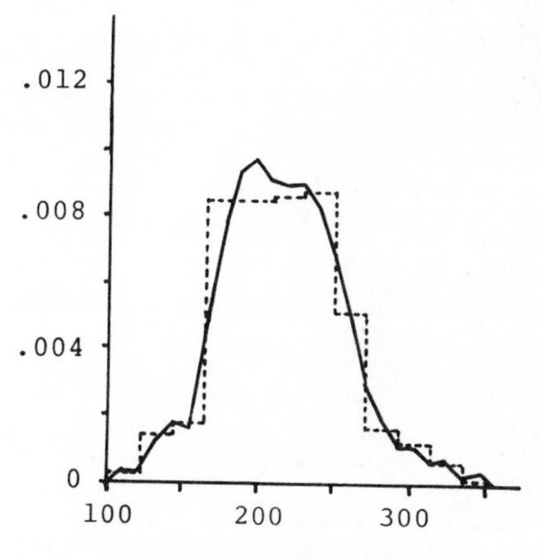

FIGURE 1B.

Plasma triglyceride concentration

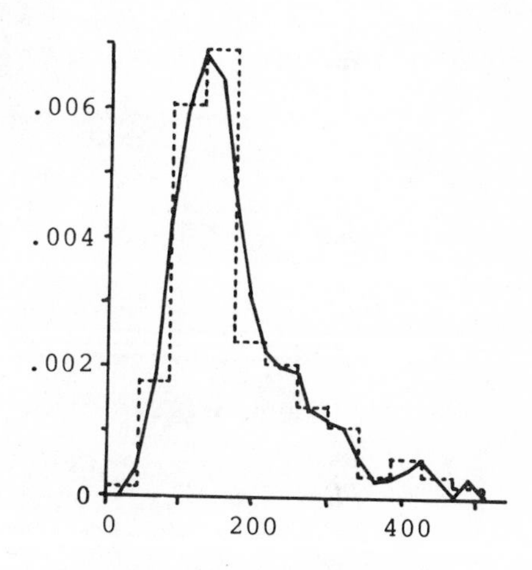

covariance matrix. The advantages of the parametric approach are ease and quality of estimation. The disadvantage is the danger of being "locked into" the wrong parametric form. In such cases information may be lost or misinterpreted.

On the other hand, the nonparametric approach attempts to approximate the "true" underlying probability density function for any parametric form. The histogram is the classical nonparametric estimator, dating back to the 17th Century. In the past twenty years three new nonparametric estimates have been investigated: kernel estimates (Parzen, 1962; Rosenblatt, 1956), series expansions (Watson and Leadbetter, 1963), and maximum penalized likelihood estimates (Good and Gaskins, 1972; deMontricher et al., 1975; Scott, 1976), the last of which is discussed in this paper. The advantage of the nonparametric approach is that these estimates are asymptotically consistent for arbitrary "true" pdf's and, thus, maximum information is preserved. The price paid for this benefit involves not only increased computational expense, but also the requirement of certain subjective decisions by the investigator. This decision-making process is not totally a disadvantage, however, since it can alert the investigator to any unusual information or features of the data.

To provide a concrete example of the amounts of information preserved in different probability density estimates, we consider a two-dimensional data set consisting of plasma cholesterol and plasma triglyceride measurements on 320 males with demonstrated occlusion of the coronary arteries. In Figures 1A and 1B we superimpose the histogram and DMPLE (see section III) estimates of the cholesterol and triglyceride measurements separately. In Gotto et al. (1977), this group of 320 males was compared to a group of males without heart disease. Univariate means, variances, and correlations of plasma lipids with extent of heart disease were presented. Only small mean differences were found. What information was lost by not graphing the pdf's in Figure 1? The heavy tail on the triglyceride graph certainly is of some interest. What further information was lost by not presenting the joint pdf of cholesterol and triglyceride? In Figure 2 we present a kernel estimate of this joint pdf by drawing contours of equal probability. This estimate is bimodal, although both marginal estimates are unimodal. In fact, the left mode in Figure 2 is indistinguishable from the single mode of the pdf calculated for the nondiseased males (Scott et al., 1978a). The speculative implication is that two mechanisms are present in heart disease. Notice the chain of information to be gleaned as we proceed up the hierarchy of estimates.

III. THE DISCRETE MAXIMUM PENALIZED LIKELIHOOD ESTIMATOR

In this section we briefly discuss the univariate nonparametric probability density estimator known as the discrete maximum penalized likelihood estimator (DMPLE) and some of its properties (Scott, 1976; Scott et al., 1978b). These estimates are either simple or piecewise linear functions on an equally spaced mesh as shown in Figure 3. The estimate is completely determined by its values at the mesh nodes. Given n data points $x_1, \ldots, x_n$, the DMPLE solves

FIGURE 2. Kernel estimate of the joint probability density function of plasma cholesterol and plasma triglyceride concentrations for 320 males with coronary artery disease. The joint density function is represented by contours of equal probability.

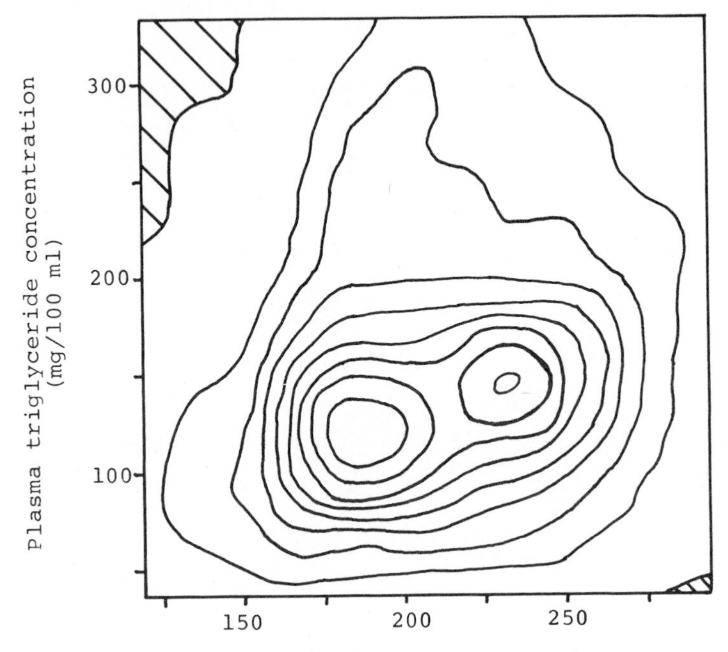

FIGURE 3. Discrete maximum penalized likelihood estimates are either simple functions s or continuous piecewise linear functions p defined on a finite interval (a,b), which is divided into an equally spaced mesh t_k.

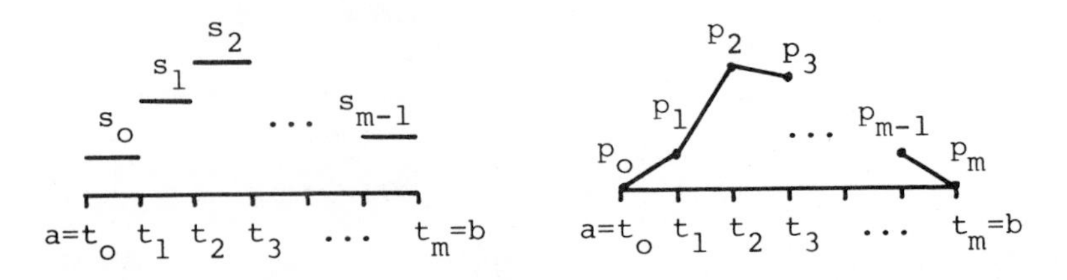

the following constrained optimization problem:

$$\text{maximize} \quad \sum_{i=1}^{n} \log s(X_i) - ch \sum \left(\frac{s_k - s_{k-1}}{h}\right)^2 \qquad (1)$$

$$\text{subject to} \quad s_k \geq 0 \quad \text{and} \quad \int s(t)dt = 1$$

where s denotes a simple function with mesh width $h = t_k - t_{k-1}$ and c is a positive constant. The first term in (1) is exactly the log of the maximum likelihood criterion and the second is a penalty term. The penalty term has the appearance of the integral of the first derivative squared of the function s and is referred to as a smoothness penalty function. Good and Gaskins (1972) first proposed the penalized maximum likelihood criterion for continuous functions; however, exact solutions for their optimization problem could not be obtained by their methods. We have chosen to find exact solutions to an approximation of their criterion, rather than to find approximate solutions (which may not integrate to one or be positive) to an exact criterion. Moreover, some of the theoretical properties of the solution to problem (1) are known. We have the following consistency result for the DMPLE: if $X_1,\ldots,X_n$ is a random sample from a density f truncated to the interval (a,b) and if f is continuous at t, then s(t) converges almost surely to the truncated density at t as h approaches zero. In fact, h may be arbitrarily small for any finite sample size without approaching Dirac spikes. Precisely, the solution to problem (1) for a fixed sample $x_1,\ldots,x_n$ converges in the sup norm as h approaches 0 to a certain polynomial spline with knots at the sample points (Scott et al., 1978b).

It is instructive to compare the roles played by c, h, a,

and b for the DMPLE with the role played by the kernel width parameter h for kernel estimates. In the latter as h approaches 0 for a fixed sample, the kernel estimates converge to Dirac spikes at the sample points. For large values of h the kernel estimate approaches zero everywhere whereas the DMPLE is constrained to a finite interval (a,b). The role of the DMPLE penalty weighing factor c is actually more analogous to the kernel parameter h. For a large weight c on the penalty term, the DMPLE is overly smoothed, while for extremely small values of c, the estimate tries to approach Dirac spikes (although the positive mesh width prevents this). Finally, prior information about the location of probability mass may be explicitly enforced in the DMPLE by choice of the support interval (a,b). Excepting the histogram, no other nonparametric estimate incorporates -- or is likely to satisfy -- such prior knowledge.

Thus the interaction of the four DMPLE parameters c, h, a, and b provides a robustness not found in other nonparametric estimates. Simulation studies have suggested the following approach for choosing the DMPLE parameters. The sample is sorted and scaled to have variance equal to one. Outlying samples are truncated to avoid unnecessary computational expense. The support interval (a,b) is chosen slightly larger than the truncated sample range. The mesh width h is chosen so that there will be about 25 mesh intervals. The choice c=1 has worked well in simulation studies of multimodal as well as unimodal data. Smaller c's may be investigated and then h reduced as desired. The resulting probability density estimate is then rescaled.

In practice the piecewise linear DMPLE with a second derivative-like penalty function is chosen for its continuity and smoothness properties. A computer library subroutine NDMPLE (IMSL, 1978) written by the authors was used to calculate all the DMPLE estimates which appear in this paper. NDMPLE uses a modified Newton's method (Tapia, 1977) and a bootstrap algorithm to avoid problems in obtaining initial values for the density estimates. For 30 mesh intervals and 400 sample points, less than 3 CPU seconds were required to obtain a DMPLE estimate on a DEC PDP-10 computer. More than a minute of CPU time was required to evaluate the kernel estimate in Figure 2 at 6800 points. Since only a linear interpolation is required to evaluate the DMPLE estimate, it is much faster to evaluate than a kernel estimate.

IV. MULTIVARIATE EXTENSION OF THE DMPLE

We propose the following constrained optimization criterion for estimating a bivariate probability density that is the simple function s_{ij} on a rectangular grid t_{ij}:

$$\max \Sigma \log s(X_i) - ch^2 \Sigma\Sigma \left((s_{ij-1}-s_{ij})/h\right)^2 + \left((s_{i-1j}-s_{ij})/h\right)^2$$

To use Newton's method we must calculate the Jacobian of this problem. It is easy to see that the Jacobian is a banded matrix and thus has a special form which will allow fast inversion. The univariate DMPLE also has a banded Jacobian matrix.

V. PSEUDO-INDEPENDENCE AND THE DMPLE

A natural question is whether a multivariate problem can be approximated by a sequence of one-dimensional problems.

In this section we investigate the possibility that a k-dimensional probability density function may be approximated by the product of k univariate probability densities. The speed of this approximation makes it worthy of investigation. One such approximation is the so-called pseudo-independence algorithm, used by Bennett (1975). It is modeled after the multivariate Gaussian feature that the k-dimensional Gaussian pdf is the product of the k univariate marginal densities when the covariance matrix is diagonal, since in this case the k random variables are independent. Given a sample of multivariate data in an n by k matrix X, the pseudo-independence algorithm is as follows: (1) Calculate the estimates M and S of the k by 1 mean vector and the k by k covariance matrix (assumed positive definite). (2) Let E be the k by k diagonal matrix containing the (positive) eigenvalues $e_1,\ldots,e_k$ of S and let V be the k by k matrix of eigenvectors. (3) Define the orthogonal transformation matrix T=V. (4) Transform the data matrix X into the n by k data matrix Y=(X-M)T, where X-M represents the mean vector M subtracted from each row of X. It is easy to see that the sample data Y has mean zero and covariance matrix given by the diagonal matrix E (pseudo-independent). (5) Estimate the k univariate probability densities $g_1,\ldots,g_k$ of the columns of Y. Then the pseudo-independent estimate of the multivariate density at a point x is given by $f(x_1,\ldots,x_k) = g_1(y_1)\ g_2(y_2)\ \ldots\ g_k(y_k)$, where y = (x-M)T and the Jacobian is equal to one.

The pseudo-independence transformation corresponds to an orthogonal axes rotation and is used in such statistical

FIGURE 4. Pseudo-independent estimate of the joint probability density function of plasma cholesterol and plasma triglyceride concentrations for 320 males with coronary artery disease. Two continuous piecewise linear discrete maximum penalized likelihood estimates with 45 mesh intervals were used to calculate the bivariate pseudo-independent estimate. The joint density function is represented by contours of equal probability. The diagonal lines represent the value zero.

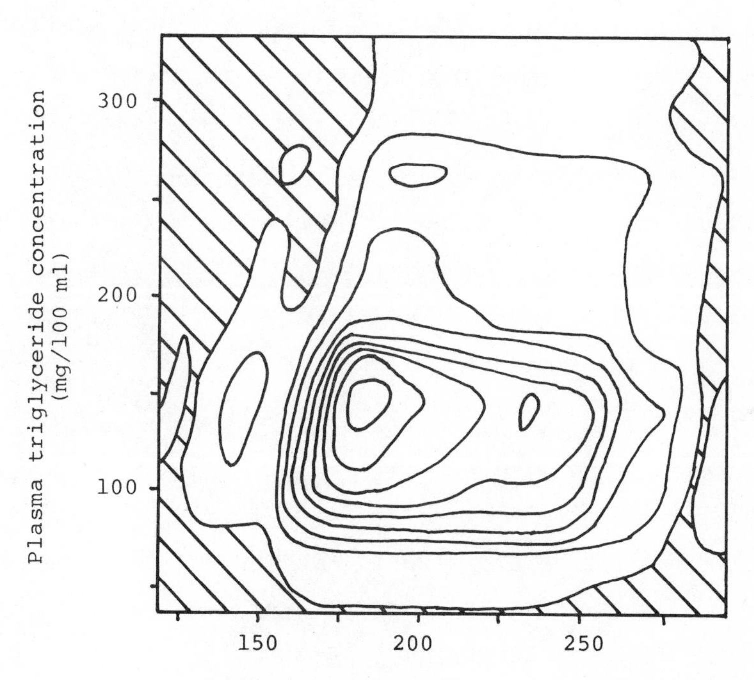

methodologies as principal components and linear discrimination. We briefly investigate the pseudo-independent probability density estimate on multimodal bivariate data. We compare this estimate's goodness-of-fit to bivariate Gaussian and histogram estimates. The goodness-of-fit measure we use is the integrated mean square error (IMSE). The IMSE measures the integrated square error between the "true" sampling pdf and the estimated pdf. The true sampling pdf's are bivariate Gaussian with various correlations or equally weighted mixtures of standardized bivariate Gaussians. Given computer-generated data, we use the DMPLE to estimate the two pseudo-independent univariate pdf's required and numerical integration to calculate the IMSE. The results of these Monte Carlo simulations are presented in Table I. We see that slightly correlated and multimodal data are well estimated using the pseudo-independence algorithm. However, the bivariate Gaussian fit is poor when the "true" pdf is not Gaussian.

Table I. Estimation Error of 3 PDF Estimators by Monte Carlo Simulation: The Average of the Integrated Mean Square Error for 10 Realizations

"True" pdf	Sample size	Number of samples	Estimator		
			Gaussian	Histogram[a]	DMPLE
r= 0[b]	100	10	.0144	.0165	.0050
r=.6[b]	100	10	.0227	.0255	.0216
r=.9[b]	100	10	.0598	.0668	.0850
mix 2[c]	200	10	.0204	.0073	.0017
mix 3[d]	300	10	.0115	.0049	.0032

[a] The histogram cells were unit squares.
[b] A standard bivariate Gaussian with given correlation r.
[c] .5N[(1.5,1.5),I] + .5N[(-1.5,-1.5),I]
[d] .33N[(1.5,1.5),I] + .33N[(-1.5,1.5),I] + .33N[(0,-1),I]

We may interpret these results by noting that the pseudo-independent data have one dimension that has maximum possible variance (using the principal components analysis). Presumably, this dimension displays the multimodal property, since projecting the modes onto one dimension in such a manner that the modes are maximally separated generally results in the maximum variance. However, when the pseudo-independence estimate was obtained for the lipid data (see Figure 4), the right mode was not as clearly demonstrated as the right mode in Figure 2 due to the heavy tails of the joint pdf. In conclusion, while there is no perfect substitute for direct estimation in higher dimensions, the pseudo-independence algorithm offers an easy, quick, and fairly accurate look at multivariate data.

ACKNOWLEDGMENT

The authors would like to thank James R. Cole, M.D. who collected the lipid data represented in the figures. Dr. Cole's present address is V.A. Hospital, Lexington, Ky.

REFERENCES

Bennett, J. (1974). Estimation of multivariate probability density functions using B-splines. Ph.D. thesis, Rice University, Houston, Texas.

de Montricher, G., Tapia, R., and Thompson, J. (1975). Nonparametric maximum likelihood estimation of probability densities by penalty function methods. Ann. Math. Stat. 3:1329.

Good, I., and Gaskins, R. (1972). Global nonparametric estimation of probability densities. Virginia J. Sci. 23:171.

Gotto, A., Gorry, G., Thompson, J., Cole, J., Trost, R., Yeshurun, and DeBakey, M. (1977). Relationship between plasma lipid concentrations and coronary artery disease in 496 patients. Cir. Nov:75.

IMSL. (1978). International Mathematical and Statistical Libraries. Houston, Texas.

Parzen, E. (1962). On estimation of a probability density function. Ann. Math. Statist. 33:1065.
Rosenblatt, M. (1956). Remarks on some nonparametric estimates of a density function. Ann. Math. Statist. 27:832.
Scott, D. (1976). Nonparametric probability density estimation by optimization theoretic techniques. TR No. 476-131-1, Rice University, Houston, Texas.
Scott, D., Gotto, A., Cole, J., and Gorry, G. (1978a). Plasma lipids as collateral risk factors in coronary artery disease: A study of 371 males with chest pain. J. Chronic Dis., to appear.
Scott, D., Tapia, R., and Thompson, J. (1978b). Nonparametric probability density estimation by discrete maximum penalized likelihood criteria. Submitted.
Tapia, R. (1977). Diagonalized multiplier methods and quasi-Newton methods for constrained optimization. J. Opt. Th. and Appl. 14:453.
Watson, G., and Leadbetter, M. (1963). On estimation of the probability density. Ann. Math. Statist. 40:1661.

SOME APPLICATIONS OF THE "CHERNOFF FACES": A TECHNIQUE FOR GRAPHICALLY REPRESENTING MULTIVARIATE DATA

Gary C. McDonald
James A. Ayers

Mathematics Department
General Motors Research Laboratories
Warren, Michigan

This paper presents a brief description and several applications of a relatively new method of graphical representation of multivariate data. The technique has been developed by H. Chernoff, and consists of mapping a vector-valued data point (presently limited to 18 or less components) into a geometrically constructed face. To provide an example of this technique the data analyzed in a recent mortality and pollution study have been mapped into faces. Each of the sixty faces represents a portrait of a particular Standard Metropolitan Statistical Area (SMSA). These faces are then employed to initialize a cluster analysis algorithm, and to examine certain trends in least squares residuals.

1. INTRODUCTION

Chernoff (1973) describes a method for mapping multivariate data into a two-dimensional cartoon face. As such, this technique provides the data analyst with a potentially very helpful tool; namely, that of a multivariate histogram which can be used to highlight important aspects or relations in a given data set. Advantages of such a graphic method may include:

(i) Serving as a convenient communicating device with respect to data description.

(ii) Detecting possible outliers or "wild" points in a data set.

ISBN 0-12-734750-X

(iii) Providing the basis for an "informal" clustering method, which may provide input into more sophisticated clustering techniques. For example, the faces may provide the experimenter with a good indication of the number of clusters and initial cluster centers to specify in other programs.

(iv) Serving as an inexpensive method of high dimensional residual analysis.

Several examples of this technique are available to the interested reader. In the Chernoff paper two examples are given. The first applies to fossil data where eight measurements were made on each of eighty-eight nummulited specimens from the Eocene Yellow Limestone Formation of northwestern Jamaica. The second example is mineral analysis data from a 4,500 foot core drilled from a Colorado mountainside which yielded twelve variables on each of fifty-three equally spaced specimens. This latter example formed the basis for a cover story in the Chemical and Engineering News, (August 28, 1972).

The purpose of this paper is to specifically illustrate the potential advantage of this technique with a particular data set involving sixteen variables and sixty observations.

2. A POLLUTION AND MORTALITY EXAMPLE DATA DESCRIPTION

In order to illustrate these applications of the faces, the data in Table 1 have been used to construct sixty faces. For a complete description of these variables and their sources, the reader is referred to

McDonald and Schwing (1973). The faces are given in Figure 1 and ordered alphabetically. Each face can be identified by the associated four letter code which serves as an abbreviation to the particular Standard Metropolitan Statistical Area (SMSA) listed in Table 1.

A characteristic of the face technique which can be exploited is that certain variables control specific features of the face. Thus, attention can be directed to the mouth (i.e., the three variables controlling the position and shape of the mouth), to the nose (i.e., the variable controlling the length of the nose), etc., or to the global impression of the facial characteristics. The variables in this data set may be separated into four convenient classes. Each class then more or less contols certain features of the faces, as noted below:

Class	Variable(s)	Features Controlled
Mortality	Total Mortality Rate	Nose length.
Climate	Precipitation	Circumferential shape of face.
	January Temperature	
	July Temperature	
	Relative Humidity	
Socioeconomic	% $\geq$ 65 years	The position and shape of the eyes, pupils and brows.
	Pop./Household	
	Education	
	% Sound Housing	
	Pop./Mile^2	
	% Non-White	
	% White Collar	
	% with income < $3000	
Pollution	HC Potential	The position and shape of mouth.
	NO_x Potential	
	SO_2 Potential	

TABLE 1. Data Analyzed by McDonald and Schwing (1973)

SMSA	PRECIPITATION	JAN. TEMP.	JULY TEMP.	% ≥ 65 YEARS	POP./HOUSEHOLD	EDUCATION
Akron, Ohio	36	27	71	8.1	3.34	11.4
Albany, N.Y.	35	23	72	11.1	3.14	11.0
Allentown, Pa.	44	29	74	10.4	3.21	9.8
Atlanta, Ga.	47	45	79	6.5	3.41	11.1
Baltimore, Md.	43	35	77	7.6	3.44	9.6
Birmingham, Ala.	53	45	80	7.7	3.45	10.2
Boston, Mass.	43	30	74	10.9	3.23	12.1
Bridgeport, Conn.	45	30	73	9.3	3.29	10.6
Buffalo, N.Y.	36	24	70	9.0	3.31	10.5
Canton, Ohio	36	27	72	9.5	3.36	10.7
Chattanooga, Tenn.	52	42	79	7.7	3.39	9.6
Chicago, Ill.	33	26	76	8.6	3.20	10.9
Cincinnati, Ohio	40	34	77	9.2	3.21	10.2
Cleveland, Ohio	35	28	71	8.8	3.29	11.1
Columbus, Ohio	37	31	75	8.0	3.26	11.9
Dallas, Tex.	35	46	85	7.1	3.22	11.8
Dayton, Ohio	36	30	75	7.5	3.35	11.4
Denver, Colo.	15	30	73	8.2	3.15	12.2
Detroit, Mich.	31	27	74	7.2	3.44	10.8
Flint, Mich.	30	24	72	6.5	3.53	10.8
Fort Worth, Tex	31	45	85	7.3	3.22	11.4
Grand Rapids, Mich.	31	24	72	9.0	3.37	10.9
Greensboro, N.C.	42	40	77	6.1	3.45	10.4
Hartford, Conn.	43	27	72	9.0	3.25	11.5
Houston, Tex.	46	55	84	5.6	3.35	11.4
Indianapolis, Ind.	39	29	75	8.7	3.23	11.4
Kansas City, Mo.	35	31	81	9.2	3.10	12.0
Lancaster, Pa.	43	32	74	10.1	3.38	9.5
Los Angeles, Calif.	11	53	68	9.2	2.99	12.1
Louisville, Ky.	30	35	71	8.3	3.37	9.9
Memphis, Tenn.	50	42	82	7.3	3.49	10.4
Miami, Fla.	60	67	82	10.0	2.98	11.5
Milwaukee, Wisc.	30	20	69	8.8	3.26	11.1
Minneapolis, Minn.	25	12	73	9.2	3.28	12.1
Nashville, Tenn.	45	40	80	8.3	3.32	10.1
New Haven, Conn.	46	30	72	10.2	3.16	11.3
New Orleans, La.	54	54	81	7.4	3.36	9.7
New York, N.Y.	42	33	77	9.7	3.03	10.7
Philadelphia, Pa.	42	32	76	9.1	3.32	10.5
Pittsburgh, Pa.	36	29	72	9.5	3.32	10.6
Portland, Ore.	37	38	67	11.3	2.99	12.0
Providence, R.I.	42	29	72	10.7	3.19	10.1
Reading, Pa.	41	33	77	11.2	3.08	9.6
Richmond, Va.	44	39	78	8.2	3.32	11.0
Rochester, N.Y.	32	25	72	10.9	3.21	11.1
St. Louis, Mo.	34	32	79	9.3	3.23	9.7
San Diego, Calif.	10	55	70	7.3	3.11	12.1
San Francisco, Calif.	18	48	63	9.2	2.92	12.2
San Jose, Calif.	13	49	68	7.0	3.36	12.2
Seattle, Wash.	35	40	64	9.6	3.02	12.2
Springfield, Mass.	45	28	74	10.6	3.21	11.1
Syracuse, N.Y.	38	24	72	9.8	3.34	11.4
Toledo, Ohio	31	26	73	9.3	3.22	10.7
Utica, N.Y.	40	23	71	11.3	3.28	10.3
Washington, D.C.	41	37	78	6.2	3.25	12.3
Wichita, Kans.	28	32	81	7.0	3.27	12.1
Wilmington, Del.	45	33	76	7.7	3.39	11.3
Worcester, Mass.	45	24	70	11.8	3.25	11.1
York, Pa.	42	33	76	9.7	3.22	9.0
Youngstown, Ohio	38	28	72	8.9	3.48	10.7

% SOUND HOUSING	POP./MILE²	% NON-WHITE	% WHITE COLLAR	% < $3000	HC POTENTIAL	NO_x POTENTIAL	SO_2 POTENTIAL	REL. HUMIDITY	TOTAL MORTALITY RATE
81.5	3,243	8.8	42.6	11.7	21	15	59	59	921.870
78.8	4,281	3.5	50.7	14.4	8	10	39	57	997.875
81.6	4,260	0.8	39.4	12.4	6	6	33	54	962.354
77.5	3,125	27.1	50.2	20.6	18	8	24	56	982.291
84.6	6,441	24.4	43.7	14.3	43	38	206	55	1,071.289
66.8	3,325	38.5	43.1	25.5	30	32	72	54	1,030.380
83.9	4,679	3.5	49.2	11.3	21	32	62	56	934.700
86.0	2,140	5.3	40.4	10.5	6	4	4	56	899.529
83.2	6,582	8.1	42.5	12.6	18	12	37	61	1,001.902
79.3	4,213	6.7	41.0	13.2	12	7	20	59	912.347
69.2	2,302	22.2	41.3	24.2	18	8	27	56	1,017.613
83.4	6,122	16.3	44.9	10.7	88	63	278	58	1,024.885
77.0	4,101	13.0	45.7	15.1	26	26	146	57	970.467
86.8	3,042	14.7	44.6	11.4	31	21	64	60	985.950
78.4	4,259	13.1	49.6	13.9	23	9	15	58	958.839
79.9	1,441	14.8	51.2	16.1	1	1	1	54	860.101
81.9	4,029	12.4	44.0	12.0	6	4	16	58	936.234
84.2	4,824	4.7	53.1	12.7	17	8	28	38	871.766
87.0	4,834	15.8	43.5	13.6	52	35	124	59	959.221
79.5	3,694	13.1	33.8	12.4	11	4	11	61	941.181
80.7	1,844	11.5	48.1	18.5	1	1	1	53	891.708
82.8	3,226	5.1	45.2	12.3	5	3	10	61	871.338
71.8	2,269	22.7	41.4	19.5	8	3	5	53	971.122
87.1	2,909	7.2	51.6	9.5	7	3	10	56	887.466
79.7	2,647	21.0	46.9	17.9	6	5	1	59	952.529
78.6	4,412	15.6	46.6	13.2	13	7	33	60	968.665
78.3	3,262	12.6	48.6	13.9	7	4	4	55	919.729
79.2	3,214	2.9	43.7	12.0	11	7	32	54	844.053
90.6	4,700	7.8	48.9	12.3	648	319	130	47	861.833
77.4	4,474	13.1	42.6	17.7	38	37	193	57	989.265
72.5	3,497	36.7	43.3	26.4	15	18	34	59	1,006.490
88.6	4,657	13.5	47.3	22.4	3	1	1	60	861.439
85.4	2,934	5.8	44.0	9.4	33	23	125	64	929.150
83.1	2,095	2.0	51.9	9.8	20	11	26	58	857.622
70.3	2,682	21.0	46.1	24.1	17	14	78	56	961.009
83.2	3,327	8.8	45.3	12.2	4	3	8	58	923.234
72.8	3,172	31.4	45.5	24.2	20	17	1	62	1,113.156
83.5	7,462	11.3	48.7	12.4	41	26	108	58	994.648
87.5	6,092	17.5	45.3	13.2	29	32	161	54	1,015.023
77.6	3,437	8.1	45.5	13.8	45	59	263	56	991.290
81.5	3,387	3.6	50.3	13.5	56	21	44	73	893.991
79.5	3,508	2.2	38.8	15.7	6	4	18	56	938.500
79.9	4,843	2.7	38.6	14.1	11	11	89	54	946.185
79.9	3,768	28.6	49.5	17.5	12	9	48	53	1,025.502
82.5	4,355	5.0	46.4	10.8	7	4	18	60	874.281
76.8	5,160	17.2	45.1	15.3	31	15	68	57	953.560
88.9	3,033	5.9	51.0	14.0	144	66	20	61	839.709
87.7	4,253	13.7	51.2	12.0	311	171	86	71	911.701
90.7	2,702	3.0	51.9	9.7	105	32	3	71	790.733
82.5	3,626	5.7	54.3	10.1	20	7	20	72	899.264
82.6	1,883	3.4	41.9	12.3	5	4	20	56	904.155
78.0	4,923	3.8	50.5	11.1	8	5	25	61	950.672
81.3	3,249	9.5	43.9	13.6	11	7	25	59	972.464
73.8	1,671	2.5	47.4	13.5	5	2	11	60	912.202
89.5	5,308	25.9	59.7	10.3	65	28	102	52	967.803
81.0	3,665	7.5	51.6	13.2	4	2	1	54	823.764
82.2	3,152	12.1	47.3	10.9	14	11	42	56	1,003.502
79.8	3,678	1.0	44.8	14.0	7	3	8	56	895.696
76.2	9,699	4.8	42.2	14.5	8	8	49	54	911.817
79.8	3,451	11.7	37.5	13.0	14	13	39	58	954.442

The pollution variables were transformed logarithmically within the FACES computer program before they were mapped into their corresponding features. Other modifications of the original program were made to provide some useful input and output options, e.g., user specified labels. With respect to the pollution potentials, the position of the mouth is directly related to HC potential. That is, low HC potential is reflected by the mouth being close to the nose (e.g., see Dallas), and large HC potentials position the mouth close to the chin (e.g., see Los Angeles). Happy faces correspond to low NO_x values while high NO_x potential results in sad faces. Finally, the length of the mouth is in part related to the SO_2 term. Generally, large SO_2 potentials result in wide mouths. The length of the nose is directly proportional to the total mortality rate. By examining the construction codes in the program, similar quantitative statements can be made concerning the other facial features and variables in this study.

Extreme conditions in one or more of the variables are usually detectable from the faces, as illustrated by the face corresponding to York, Pa. The eyes clearly indicate that the socioeconomic conditions in York may be rather unique compared to other observations in our data set. A closer examination of the data reveals that York has the minimum value for education and the maximum value for population density (pop./Mile^2).

3. APPLICATION TO CLUSTER ANALYSIS

Each of the authors, as well as several colleagues, have independently and "informally" clustered these faces without knowing which SMSA's the faces represented. There did appear to be a great deal of similarity in the resulting clusters; e.g., the four California faces were clustered together. Cleveland, Milwaukee, Boston, Chicago, Louisville, Pittsburgh,

FIGURE 1. Faces Ordered Alphabetically

and Detroit were considered similar. The SMSA's of Kansas City, Wichita, Fort Worth, Dallas and Greensboro were thought to be "related." Although agreement between all clusters was not perfect, the faces did provide a quick and inexpensive "first look" at the data, and does suggest that further development of these methods may be beneficial.

One such clustering of these faces resulted in four main groups of SMSA's as follows:

I. Los Angeles, San Diego, San Francisco, San Jose;

II. Dallas, Fort Worth, Greensboro, Kansas City, Wichita;

III. Atlanta, Baltimore, Birmingham, Chattanooga, Cincinnati, Columbus, Houston, Memphis, Miami, Minneapolis, Nashville, New Orleans, New York, Philadelphia, Reading, Richmond, St. Louis, Washington D.C., Wilmington, York;

IV. Akron, Albany, Allentown, Boston, Bridgeport, Buffalo, Canton, Chicago, Cleveland, Dayton, Denver, Detroit, Flint, Grand Rapids, Hartford, Indianapolis, Lancaster, Louisville, Milwaukee, New Haven, Pittsburgh, Portland, Providence, Rochester, Seattle, Springfield, Syracuse, Toledo, Utica, Worcester, Youngstown.

These clusters consist of 4, 5, 20 and 31 members, respectively.

This information was then considered "input" to the algorithm provided by D. N. Sparks (1973). This algorithm performs the essential calculations of Euclidean cluster analysis described by E. M. L. Beale (1969), and requires for input: (i) the matrix of observations, (ii) the number of clusters, (iii) the initial cluster centers, and (iv) the minimum number of observations which any cluster is allowed to have. The matrix of observations was taken to be Table 1 with each entry standardized by subtracting the mean and then dividing by the standard deviation of the variable based on the sixty observations. The number of clusters was set

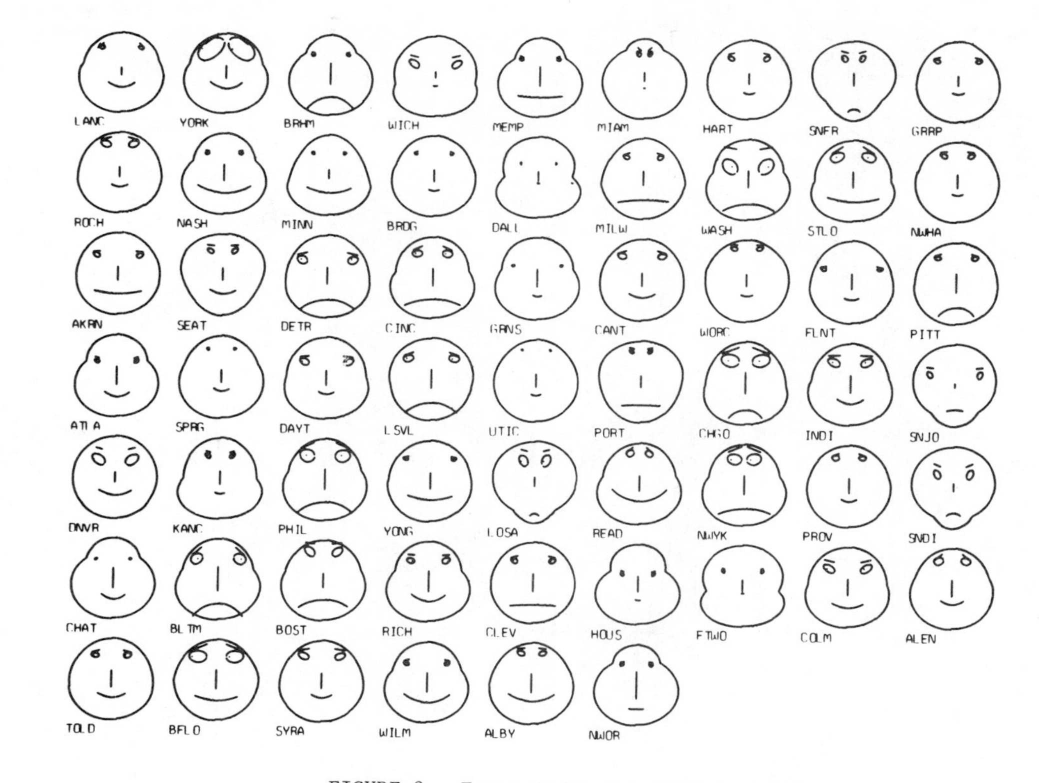

FIGURE 2. Faces Ordered by OLS Residuals

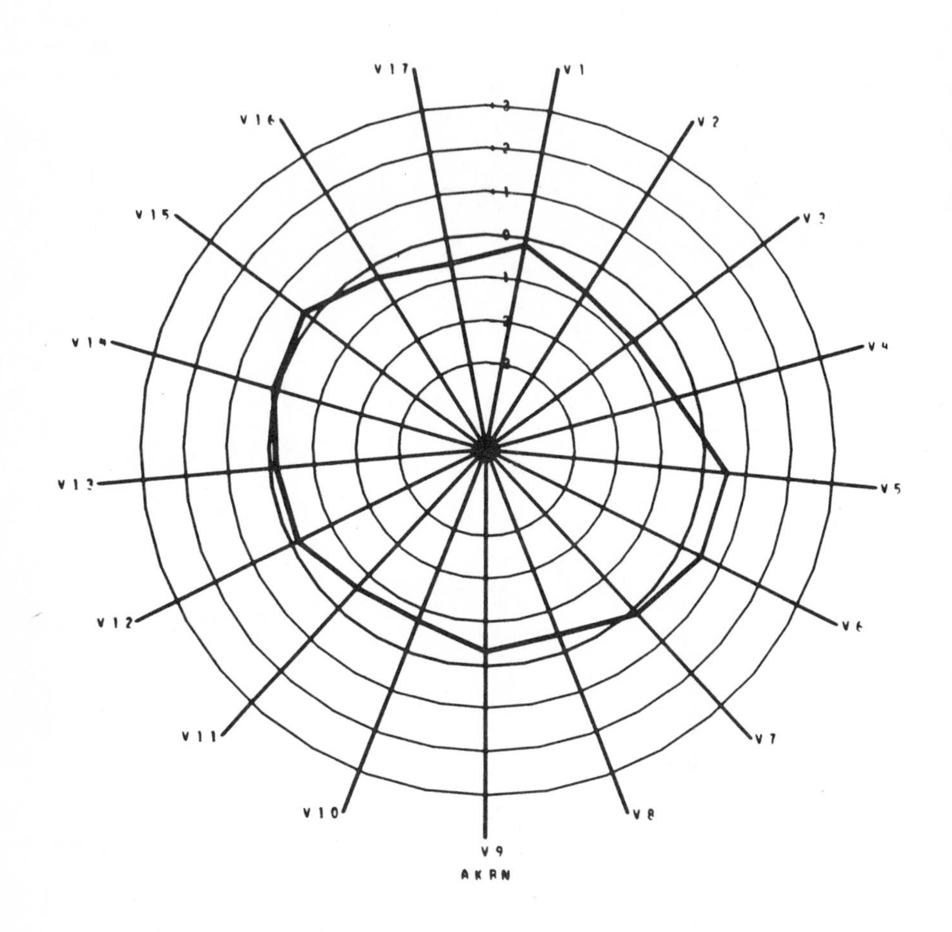

FIGURE 3. Circle Diagram for Akron

at four, the initial cluster centers were computed from the original clustering given above and each cluster was allowed no less than four observations. The results of this analytical clustering scheme were:

I. Los Angeles, San Diego, San Francisco, San Jose;

II. Dallas, Denver, Fort Worth, Kansas City, Miami, Washington D.C., Wichita;

III. Atlanta, Birmingham, Chattanooga, Greensboro, Houston, Memphis, Nashville, New Orleans, Richmond;

IV. Akron, Albany, Allentown, Baltimore, Boston, Bridgeport, Buffalo, Canton, Chicago, Cincinnati, Cleveland, Columbus, Dayton, Detroit, Flint, Grand Rapids, Hartford, Indianapolis, Lancaster, Louisville, Milwaukee, Minneapolis, New Haven, New York, Philadelphia, Pittsburgh, Portland, Providence, Reading, Rochester, St. Louis, Seattle, Springfield, Syracuse, Toledo, Utica, Wilmington, Worcester, York, Youngstown.

These Clusters now have 4, 7, 9 and 40 members, respectively; and the reallocation among the four clusters is indicated in Table 2. Forty-six of the sixty cases remained in their original clusters while fourteen cases were displaced.

TABLE 2. Reallocation Matrix of Four Clusters

		Final Clusters				
		I	II	III	IV	Total
Original Clusters	I	4	0	0	0	4
	II	0	4	1	0	5
	III	0	2	8	10	20
	IV	0	1	0	30	31
Total		4	7	9	40	60

FIGURE 4. Snowflakes Ordered Alphabetically

4. APPLICATION TO RESIDUAL ANALYSIS

In order to illustrate an application of these faces to residual analysis, the sixty faces of the mortality and pollution data given in Figure 1 have been ordered in Figure 2 in increasing order of the residual. The residual for a given Standard Metropolitan Statistical Area is the actual mortality rate minus the predicted rate based on an ordinary least squares (OLS) regression equation with fifteen explanatory variables. As indicated, Lancaster, Pa. has the smallest residual, York, Pa. the next smallest,..., and New Orleans, La. has the largest residual in this analysis. These ordered faces can now be visually examined for trends in one or more of the facial features. It should be noted that facial features (e.g., mouth shape, facial outline, etc.) are typically functions of several variables, and such trend patterns (if they exist) could easily be unobservable in the plots of residuals versus each of the variables usually provided by a standard "canned" program. The labeling of the faces may also be exploited to examine other possible trends in the residuals; e.g., in this particular instance, the labeling provides information on the geographic distribution of the residuals. An examination of Figure 2 reveals no drastic abnormal behavior with respect to the ordered faces. However, the faces corresponding to the eight largest residuals are all "happy," which indicates a relatively low NO_x pollution potential, and thus may indicate a slight departure from some of the basic regression assumptions underlying the linear model employed. The analyst may then proceed with a more critical examination of the linearity assumptions of the variable or variables in question. Specific questions regarding the NO_x pollution potential variable have been considered by McDonald and Schwing (1973).

5. CONCLUSIONS

Graphical techniques for inexpensively displaying multivariate data are very useful in exploratory data analysis and data editing. The faces technique is one of many such techniques which have been, in some instances, usefully exploited to such ends. An inherent characteristic of the faces method is that some features of the face have a greater distinguishing effect to the analyst than do other features. This characteristic may be advantageously exploited if it is desired to highlight or emphasize certain aspects of the data. However, this characteristic may also serve to weight disproportionately a subset of the variables in a visual clustering of the observations. For example, it has been our experience that the facial outline plays a more crucial role in visual clustering than do the individual eye features. Perhaps an experienced analyst may be able to partially overcome these difficulties. However, a more promising strategy might be to permute the variables which do control the individual facial features and monitor the invariance (or lack of invariance) in the resulting visual clusters. Alternatively, one may employ a different type of graphical representation of the data. In the examples presented in this paper, no attempt has been made to ascertain a degree of invariance with respect to permuting the variables controlling specific features of the face.

A graphical technique which is not as likely to introduce a disproportionate weight on any subset of the variables as the faces method is the construction of circle diagrams, e.g., see Siegel, _et al_. (1972). The diagrams are constructed by connecting with straight line segments, the standardized variables plotted along equally spaced rays of a circle. Figure 3 is such a diagram for the Akron SMSA and the notation V1,..., V16 refers to the variables listed in Table 1 ordered from precipitation to total mortality. The variable V17 is the residual from a least squares regression fit. Upon removing the grid structure from the circle diagram, the resulting polygon appears as a "snowflake" as indicated in Figure 4. This method of data description has been utilized by Herman and Montroll (1972), and can be easily exploited for a rough visual clustering and/or residual examination as were the cartoon faces.

ACKNOWLEDGMENTS

The authors would like to gratefully acknowledge the efforts of Mr. H. Gugel in implementing an interactive circle diagram computer package on the GMR system.

REFERENCES

Beale, E. M. L. (1969). Euclidean cluster analysis. Contributed paper to the 37th session of the International Statistical Institute.

Chernoff, H. (1973). The use of faces to represent points in k-dimensional space graphically, Journal of the American Statistical Association (68), 361-368.

Herman, R. and Montroll, E. W. (1972). A manner of characterizing the development of countries, Proc. Nat. Acad. Sci. U.S.A. (69), 3019-3023.

McDonald, G. C. and Schwing, R. C. (1973). Instabilities of regression estimates relating air pollution to mortality, Technometrics (15), 463-481.

Siegel, J. H., Farrell, E. J., Goldwyn, R. M. and Friedman, H. P. (1972). The surgical implication of physiologic patterns in myocardial infarction shock, Surgery (72), 126-141.

Sparks, D. N. (1973). Euclidean cluster analysis, algorithm AS 58, Applied Statistics (22), 126-130.

A MULTIVARIATE GRAPHIC DISPLAY FOR REGIONAL ANALYSIS

David L. Huff
William Black

Graduate School of Business Administration
University of Texas
Austin, Texas

I. INTRODUCTION

As the mass of data generated to understand urban and regional problems increases both in terms of variety as well as complexity, better methods are needed to display, communicate, and analyze such information. The use of graphic displays represents an important and underutilized medium of transmitting information and for exploratory data analysis. Such displays may evoke impressions of underlying relationships that might not easily be detected be mathematical techniques. Graphic displays, therefore, serve as methodological aids to hypothesis discovery. Hypothesis discovery is quite different from hypothesis validation. This distinction is exemplified in the following passage by Newton:

> The prosess of hypothesis discovery is quite different. Though rooted in established knowledge, it also is the child of poorly understood faculties of the mind. The very riskiness of this process is what enables it to advance science--to lead beyond the boundaries of what can safely be deducted from extablished facts. It thus cannot and should not be held to the rigorous rules of objectivity that are appropriate to hypothesis validation. This need not be a matter of concern so long as one understands that hypothesis dis-

ISBN 0-12-734750-X

> covery does not stand alone, that scientific enterprise is the intelligent interplay of two processes, discovery and validation. By requiring rigor in the validation process, we can liberate the discovery process.[1]

Traditionally, graphic data displays have been limited principally to maps, charts, histograms, and scatter diagrams. While such displays do provide a simplistic and an effective means of conveying certain data features, they cannot be used to display multidimensional data, and they are limited in terms of providing a basis for generalizing and communicating the relationships that are portrayed.

II. PURPOSE

The use of computer graphics makes it possible to transform data into a number of different types of geometrical forms. The purpose of this article is to present a method of converting multivariate data into faces. The method was originally developed by Chernoff and consists of representing a point in k-dimensional space as a cartoon of a face, whose facial features correspond to components of the point.[2] The mechanics of constructing a face will be described in the section immediately following. Then, the use of faces as a classification aid will be exemplified. Finally, suggestions will be made with respect to various uses that might be made of this type of multivariate graphic display in urban and regional analysis.

---

[1] Carol M. Newton, "Graphic Data Analysis: Optimizing Tradeoffs Between Richness and Simplicity," Proceedings of the American Statistical Association, 1976.

[2] Herman Chernoff, "The Use of Faces to Represent Points in k-Dimensional Space Graphically," Journal of the American Statistical Association, Vol. 68, No. 342, 1973.

III. CONSTRUCTING THE FACE

The number of variables required to draw a face can vary depending on the number of facial variables desired. The computer program utilized in this study can employ up to twenty variables to construct a face.[3] Figure 1 and Table 1 describe the facial features controlled by the program variables X_1, $X_2, \ldots X_{20}$. For example, X_6 controls the length of the nose while X_8 controls the curvature of the mouth. Also given in Table 1 are the ranges for the program variables. These ranges are the possible values the program variables may assume after the original variable values are scaled within their upper and lower bounds.

The construction of the face involves six basic facial features: (1) head; (2) mouth; (3) nose; (4) eyes; (5) eyebrows; and, (6) ears. Table 1 lists the appropriate program variable for each facial feature. The head is composed of two ellipses intersecting at points P and P' and oriented about horizontal and vertical axes passing through the origin point 0. As seen in Figure 1, U and L represent the upper and lower vertical limits of the face. The distances OU or OL are equal and termed the half-height of the face (h). The points of intersection, P and P', are dependent on the values of OP and θ*. These two program variables combine to determine P and P', which are symmetrical about the vertical axis with respect to 0. Finally, the sizes and shapes of the ellipses are determined by the half-height of

---

[3]The computer program utilized in this study was a modified version of PROGRAM DRFACE obtained from Lawrence A. Bruckner, University of California, Los Alamos Scientific Laboratory, Los Alamos, New Mexico.

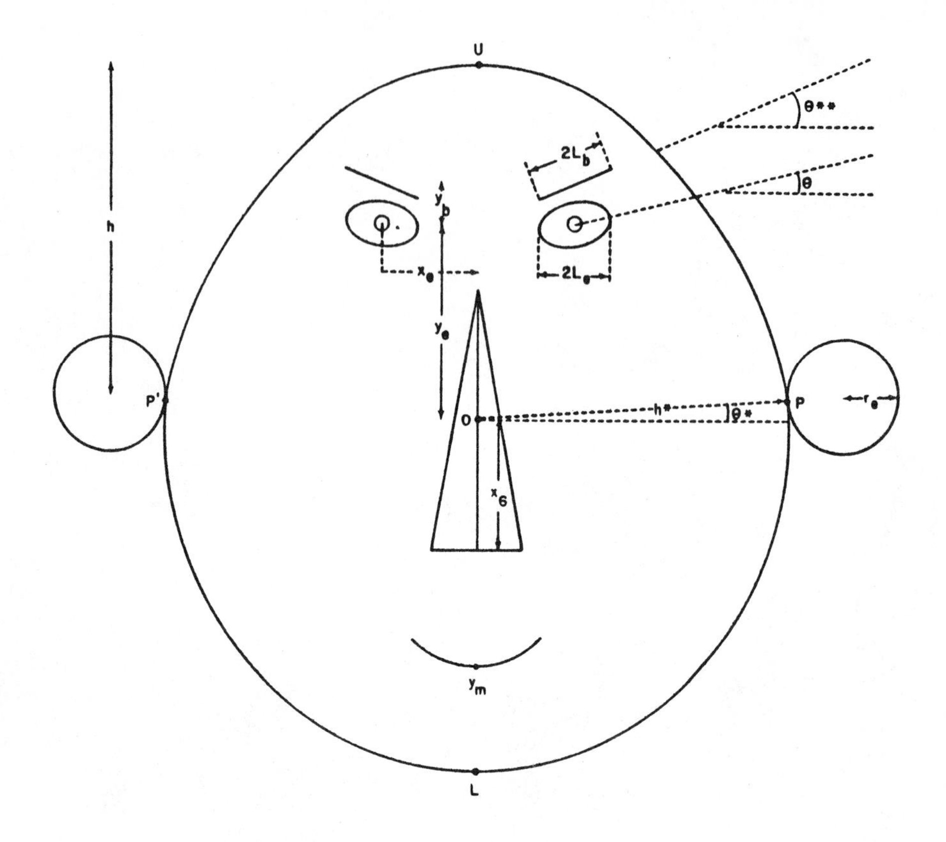

Figure 1. Basic Features in Facial Construction

the face, the intersection points and the eccentricities of each ellipse. Both ellipses have major axes coinciding with the vertical axis.

The mouth is a circular arc which is centered on the vertical axis and passes through the point Ym. Since the program variable X_8 can assume negative as well as positive values, the curvature of the mouth can vary from a frown to a smile. The nose is a triangle centered at 0, with both the height and width controlled by the program variables X_6 and X_{20}. The eyes are ellipses oriented symetrically about

TABLE I. Program Variables Employed in Constructing a Face

Variable	Range	Description	
x_1	(.2, .7)	Distance from origin 0 to P	h^*
x_2	(.35, .65)	Angle between OP and X-axis	θ^*
x_3	(.5, 1.0)	Half height of face	h
x_4	(.5, 1.0)	Eccentricity of upper ellipse	
x_5	(.5, 1.0)	Eccentricity of lower ellipse	
x_6	(.15, .4)	Length of nose	
x_7	(.2, .8)	Position of center of mouth	Y_m
x_8	(−4, 4)	Curvature of mouth	
x_9	(.3, 1)	Length of mouth	am
x_{10}	(0, .3)	Height of centers of eyes	Y_e
x_{11}	(.3, .8)	Separation of center of eyes	x_e
x_{12}	(.2, .6)	Slant of eyes	θ
x_{13}	(.4, .8)	Eccentricity of eyes	
x_{14}	(.2, 1.0)	Half-length of eyes	Le
x_{15}	(.2, .8)	Position of pupils	
x_{16}	(.6, 1.0)	Height of eyebrow center relative to eye	Y_b
x_{17}	(0, 1.0)	Angle of eyebrow	θ^{**}
x_{18}	(.3, 1.0)	Length of eyebrow	L_b
x_{20}	(.1, 1.0)	Ear diameter	Y_e
x_{21}	(.1, .2)	Nose width	

the vertical axis. The positioning (vertical and horizontal), slant, eccentricity and size, of the eyes can all be varied by the program variables X_{10}, X_{11}, X_{12}, X_{13} and X_{14}. The eyebrows are line segments which can be varied in length, slant, and vertical positioning. The final features are the ears, which are represented by circles tangent to points P and P' whose radii are determined by X_{19}.

IV. AN EXAMPLE

A. The Data

A comparison of thirteen large metropolitan areas on seven economic measures will serve to illustrate the use of faces as a potential aid in classification analysis. The selected metropolitan areas include:

Baltimore	Minneapolis-St. Paul
Chicago	New York
Cleveland	Philadelphia
Dallas	Pittsburgh
Detroit	St. Louis
Houston	San Francisco
Los Angeles	

The seven economic variables pertaining to each of these metropolitan areas are listed below:

transfer payments

change in transfer payments (1970-73)

per capita taxes

change in per capita taxes (1970-73)

unemployment

cost of living

per capita income

The data are shown in Table 2. No preference was given to the metropolitan areas selected. Data on cost of living are reported only for a limited number of metropolitan areas and, thus, the inclusion of this variable restricted the areas that could be included. The variables that were selected reflect, in general, the economic well being of an area. Other measures could also have been included but an attempt was made to limit the data set to a few of the more important measures.

B. Facial Features

As was mentioned previously, the computer program used in this study utilizes twenty variables to draw a face. The seven variables that were selected for each of the metropolitan areas were assigned to the following facial features:

Variables	Facial Features
1. per capita transfer payments	length of nose
2. change in per capita transfer payments	width of nose
3. per capita taxes	half length of eyes
4. change in per capita taxes	eccentricity of eyes
5. unemployment	angle of eyebrow
6. cost of living index	curvature of mouth
7. per capita income	diameter of ears

The assignment of variables to facial features was arbitrary as has been the practice in other studies using faces.[4]

Reciprocals were taken for per capita taxes, change in

[4]Lawrence A. Bruckner, "The Looks of Some Companies Involved in Offshore Oil and Gas Leases," University of California, Los Alamos Scientific Laboratory, Los Alamos, California, LA-UR-76-462.

TABLE II. Selected Economic Variables for Thirteen Metropolitan Areas

	Per Capita Transfer Payments[1]		Per Capita Taxes + Current Charges[2]		Unemployment[3]	Cost of Living Index[4]	Per Capita Income[5]
	(1973)	(1970-1973)	(1973)	(1970-1973)			
Baltimore	$466	$165	$331	$ 71	3.8%	134.9	$5733
Chicago	622	269	458	167	4.9	132.0	6160
Cleveland	593	201	412	66	7.1	134.1	5960
Dallas	390	199	361	117	2.4	132.0	5669
Detroit	618	213	445	112	9.1	134.5	5989
Houston	361	117	315	71	4.0	132.3	5439
Los Angeles	662	182	465	63	6.4	129.2	5757
Minneapolis-St. Paul	583	196	375	101	5.2	133.0	5992
New York	760	233	706	216	6.0	139.7	6117
Philadelphia	812	290	385	74	6.8	135.5	4925
Pittsburgh	663	231	289	15	5.7	132.9	5295
St. Louis	480	165	288	40	5.6	129.3	5653
San Francisco	816	237	590	65	10.6	131.5	6210

[1] U.S. Department of Commerce, Bureau of Economic Analysis, Personal Income by Major Sources; 1970-1975, Washington, D.C. (1975).

[2] U.S. Department of Commerce, Bureau of the Census, Local Government Finances in Selected Metropolitan Areas and Large Counties: 1973-74; Series GF-74, No. 6, and, Local Government Finances in Selected Metropolitan Areas and Large Counties: 1970-1971, Series GF-71, No. 6.

[3] U.S. Department of Labor, Bureau of Labor Statistics, Handbook of Labor Statistics: 1975, Reference Edition, Table 52, pp. 127-129; and, Handbook of Labor Statistics, 1972, Table 51, pp. 101-103.

[4] U.S. Department of Labor, Bureau of Labor Statistics, Handbook of Labor Statistics: 1976, Bulletin 1905, Table 122, pp. 257-259.

[5] U.S. Department of Commerce, Bureau of Economic Analysis, Personal Income by Major Sources; 1970-1975, Washington, D.C. (1975).

per capita taxes, unemployment, and the cost of living indices. As a consequence, the face of a metropolitan area that has low per capita transfer payments would have a short nose; the width of the nose would be narrow if the change in per capita transfer payments were small; if the reciprocal of per capita taxes were large, a face would have large round eyes; the larger the reciprocal of unemployment, the smaller will be the angle of the eyebrows; the larger the reciprocal of the cost of living index, the more upturned will be the curvature of the mouth; finally, the higher the level of per capita income, the larger will be the ears.

C. Facial Similarities

A hierarchical cluster analysis was performed in an effort to determine the similarity of the metropolitan areas with respect to the seven variables which had been converted to standard scores (see Table 3). The smallest within group variance was obtained with five groups. The metropolitan areas within each group are listed below:

Group 1	Group 2	Group 3
St. Louis	Chicago	New York
Baltimore	Detroit	
Houston	Cleveland	
Dallas	Mpls.-St. Paul	
	San Francisco	
	Los Angeles	

Group 4	Group 5
Philadelphia	Pittsburgh

Figure 2 shows the face of each metropolitan area and the group to which it was assigned on the basis of the results obtained from the cluster analysis. It can be seen that each face is unique, yet, those within a group appear, in most cases, to bear more resemblance to one another than those in different groups.

TABLE III. Standard Scores of the Variables

	PC Transfer Payments	Change in PC Transfer Payments	PC Taxes	Change in PC Taxes	Unemploy-ment	Cost of Living Index	PC Income
Baltimore	-1.04	- .93	.71	- .17	.86	- .67	- .76
Chicago	.15	1.34	- .63	0.67	.15	.44	1.07
Cleveland	- .05	- .14	- .23	- .10	- .61	- .34	.53
Dallas	-1.41	- .19	.31	- .51	2.71	.39	- .25
Detroit	.12	.12	- .20	- .49	- .99	- .54	.61
Houston	-1.61	-1.97	.95	- .17	.70	.31	- .86
Los Angeles	.41	- .56	· .68	- .08	- .43	1.49	- .12
Minneapolis-St. Paul	- .12	- .25	.15	- .43	.00	.05	.62
New York	1.07	.55	1.85	- .76	- .30	-2.31	.95
Philadelphia	1.42	1.79	.04	- .20	- .54	- .87	-2.24
Pittsburgh	.42	.51	1.42	3.14	- .20	.05	-1.25
St. Louis	- .81	- .81	1.43	.52	- .17	1.43	- .29
San Francisco	1.45	1.45	-1.42	- .08	-1.18	.58	1.20

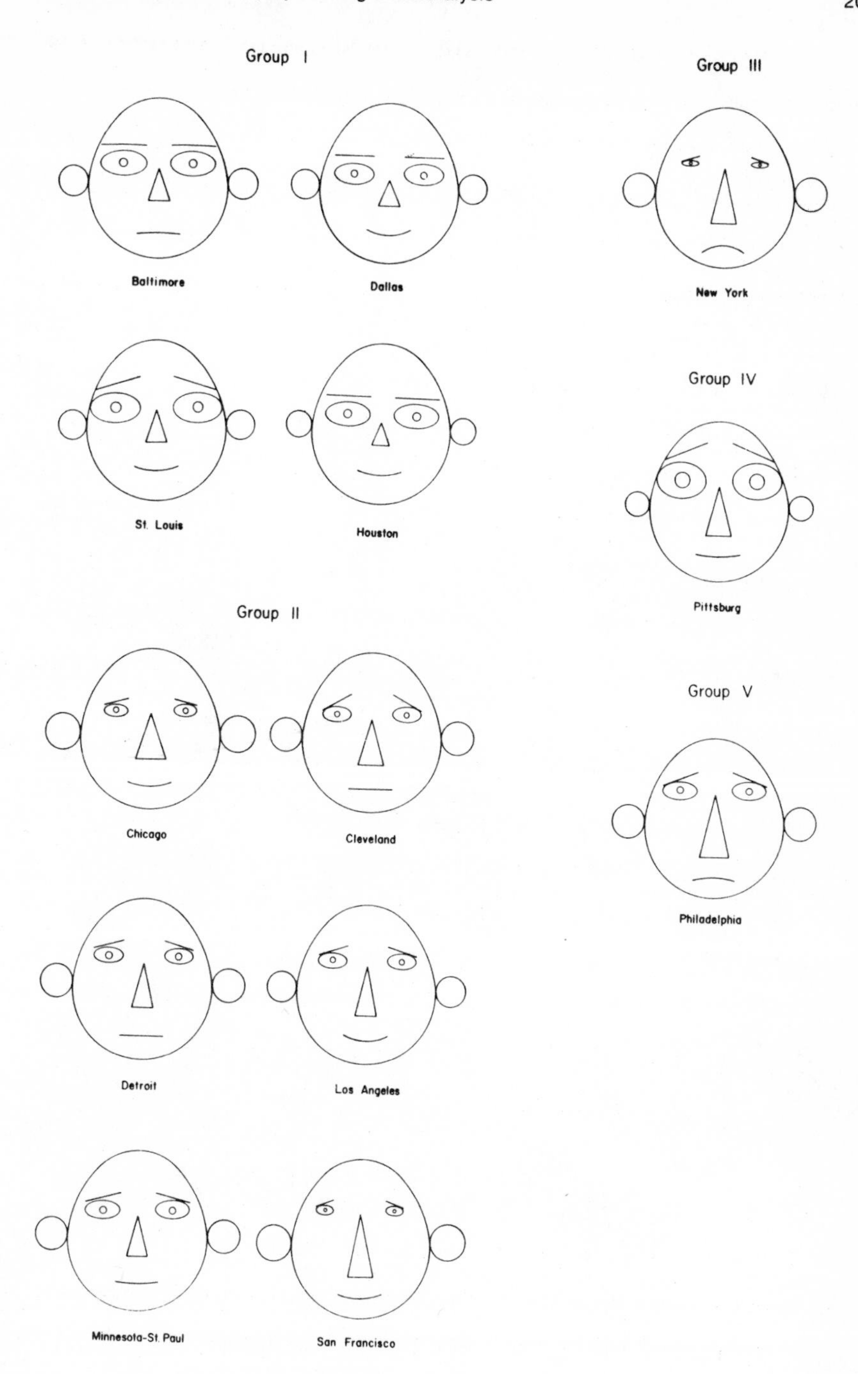

Figure 2. Faces of the Metropolitan Areas and Their Group Membership: First Experiment

One of the objectives of this study was to determine how closely human observers could duplicate a statistically derived typology based solely on perceived differences among faces. In an effort to assess this capability 40 students were asked individually to sort the faces into groups that appeared to look most similar. The number of groups was restricted to five so that they would conform to the five groups that were obtained from the cluster analysis.

The results of the matching of faces by the 40 students (actual) in comparison to the matchings based on the results of the cluster analysis (expected) are indicated in Table 4. It can be seen that, in general, there was a greater correspondence between the actual and expected pairing for the four metropolitan areas comprising Group 1. The least-correspondence occurred for the three single member groups (New York, Philadelphia, and Pittsburgh). The correlation coefficients between the actual and expected values for the five groups are indicated below.

Group 1	.82
Group 2	.58
Group 3	.05
Group 4	.13
Group 5	.36

D. The Relative Importance of Different Facial Features

The initial group of 40 students was asked, after they had completed making their facial grouping, to rank the seven facial features in terms of how important they felt each was in differentiating the faces from one another. The results of the rank ordering are shown below.

TABLE IV. A Comparison of Actual and Expected Pairings From the First Experiment

(Actual - *Expected*)

	Baltimore	Chicago	Cleveland	Dallas	Detroit	Houston	Los Angeles	Minneapolis-St. Paul	New York	Philadelphia	Pittsburgh	St. Louis	San Francisco
Baltimore	8	0	6	18	5	21	0	2	1	3	2	9	0
	0	*0*	*0*	*40*	*0*	*40*	*0*	*0*	*0*	*0*	*0*	*40*	*0*
Chicago	0	0	7	4	10	3	32	22	0	5	2	4	19
	0	*0*	*40*	*0*	*40*	*0*	*40*	*40*	*0*	*0*	*0*	*0*	*40*
Cleveland	6	7	0	0	34	1	5	6	3	19	1	0	0
	0	*40*	*0*	*0*	*40*	*0*	*40*	*40*	*0*	*0*	*0*	*0*	*40*
Dallas	18	4	0	0	0	37	3	5	0	0	4	17	2
	40	*0*	*0*	*0*	*0*	*40*	*0*	*0*	*0*	*0*	*0*	*40*	*0*
Detroit	5	10	34	0	0	0	8	9	3	14	0	0	0
	0	*40*	*40*	*0*	*0*	*0*	*40*	*40*	*0*	*0*	*0*	*0*	*40*
Houston	21	3	1	37	0	0	2	3	0	0	4	18	2
	40	*0*	*0*	*40*	*0*	*0*	*0*	*0*	*0*	*0*	*0*	*40*	*0*
Los Angeles	0	32	5	3	8	2	0	26	2	5	2	5	18
	0	*40*	*40*	*0*	*40*	*0*	*0*	*40*	*0*	*0*	*0*	*0*	*40*
Minneapolis-St. Paul	2	22	6	5	9	3	26	0	0	5	9	8	13
	0	*40*	*40*	*0*	*40*	*0*	*40*	*0*	*0*	*0*	*0*	*0*	*40*
New York	1	0	3	0	3	0	2	0	5	18	0	0	18
	0	*0*	*0*	*0*	*0*	*0*	*0*	*0*	*40*	*0*	*0*	*0*	*0*
Philadelphia	3	5	19	0	14	0	5	5	18	3	1	1	3
	0	*0*	*0*	*0*	*0*	*0*	*0*	*0*	*0*	*40*	*0*	*0*	*0*
Pittsburgh	2	2	1	4	0	4	2	9	0	1	12	21	3
	0	*0*	*0*	*0*	*0*	*0*	*0*	*0*	*0*	*0*	*40*	*0*	*0*
St. Louis	9	4	0	17	0	18	5	8	0	1	21	0	3
	40	*0*	*0*	*40*	*0*	*40*	*0*	*0*	*0*	*0*	*0*	*0*	*0*
San Francisco	0	19	0	2	0	2	18	13	18	3	3	3	0
	0	*40*	*40*	*0*	*40*	*0*	*40*	*40*	*0*	*0*	*0*	*0*	*0*

1. Half length of eyes (most important)
2. Eccentricity of eyes
3. Angle of eyebrow
4. Curvature of mouth
5. Length of nose
6. Width of nose
7. Diameter of ears (least important)

A one-way analysis of variance was then performed to determine the relative importance of each of the seven economic measures based on the amount of total variance each explained. The results are depicted in Table 5.

TABLE V. Simple Analysis of Variance of the Economic Measures

Variable Name	Univariate F-Ratio
Change in per capita taxes	24.8
Per capita income	16.7
Per capita transfer payments	9.6
Per capita taxes	9.2
Change in per capita transfer payments	4.0
Cost of living index	2.9
Unemployment	2.1

It is interesting to note the differences in the initial assignment of variables to facial features and the assignment if the student's rank order of facial features is matched to those derived from the analysis of variance. These differences are indicated below.

Rank Order Based on Analysis of Variance	Initial Assignment of Facial Features
Change in PC taxes	Eccentricity of eyes
PC income	Diameter of ears
PC transfer payments	Length of nose
PC taxes	Half length of eyes
Change in PC transfer payments	Width of nose
Cost of living index	Curvature of mouth
Unemployment	Angle of eyebrow

Student's Rank of Facial Features
Half length of eyes
Eccentricity of eyes
Angle of eyebrow
Curvature of mouth
Length of nose
Width of nose
Diameter of ears

E. Second Experiment

A second experiment was conducted in order to determine if a closer correspondence between student groupings and those produced by the cluster analysis could be obtained if the student's rank order of facial features was matched to the rank order of variables produced by the analysis of variance. The new faces generated by these changes are depicted in Figure 3. A different set of 40 students were asked to sort the redrawn faces into five groups. A comparison of the actual and expected pairing from the second

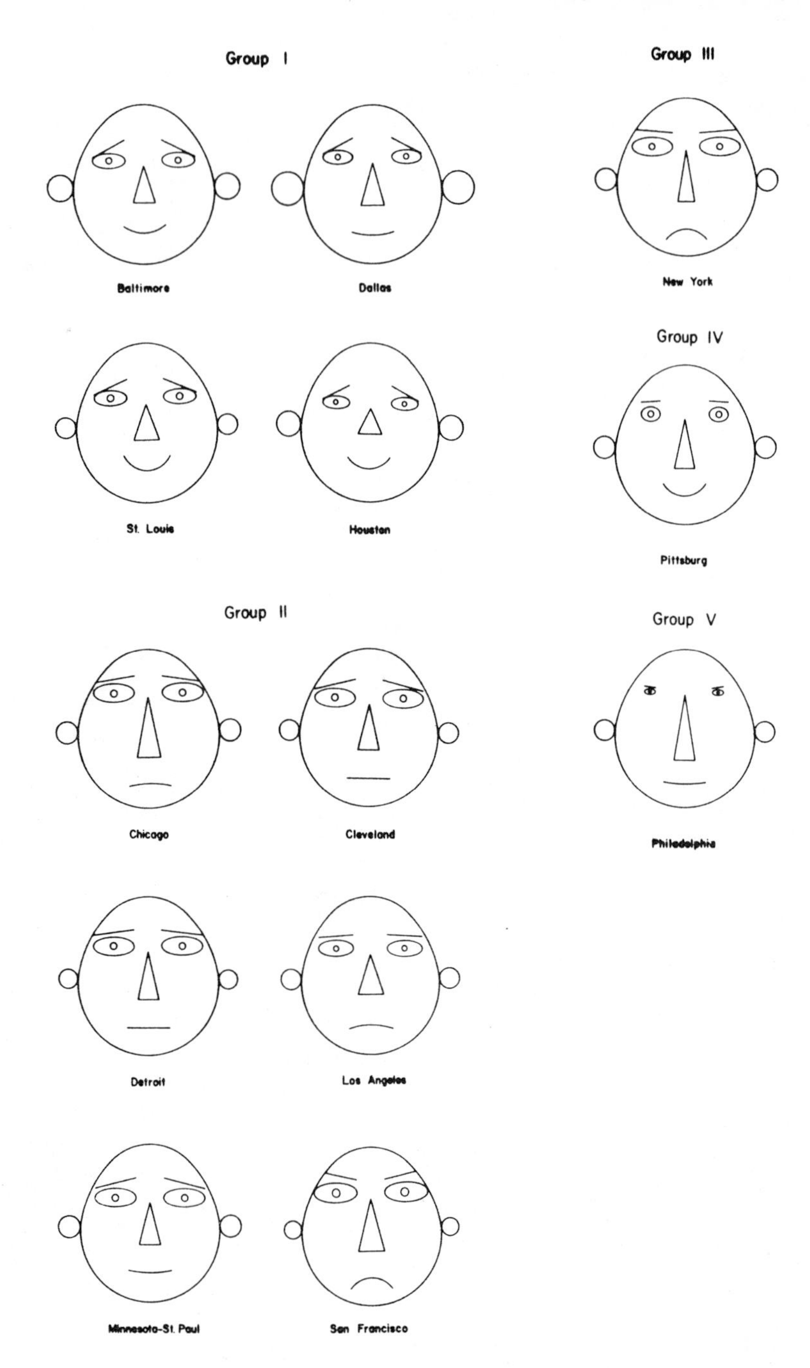

FIGURE 3. Faces of Metropolitan Areas and Their Group Membership: Second Experiment

experiment are shown in Table 6. The correspondence between the actual and expected groupings increased markedly as indicated by the correlation coefficients noted below.

	Correlation Coefficients	
	Second Experiment	First Experiment
Group 1	.94	.82
Group 2	.62	.58
Group 3	.16	.05
Group 4	.99	.13
Group 5	.64	.36

Once again, the lowest correspondence was with Group 3 (New York). Thirty-eight of the 40 students thought that the face depicting New York should be paired with the one for San Francisco to form a group. An inspection of the faces for these two metropolitan areas, as shown in Figure 3, confirms that they do appear almost identical. Yet, based on the cluster analysis, they are quite different. If a detailed comparison of the two faces is made, it can be seen that there are a number of differences despite their apparent similarity.

Two of the three single member groups were indicated by more of the students in the second experiment. Thirty-seven believed the face for Philadelphia was unique and should represent a single member group. Twenty-two students, as contrasted to 12 in the first experiment, felt the face for Pittsburgh was also unique.

TABLE VI. A Comparison of Actual and Expected Pairings From the Second Experiment

(Actual - *Expected*)

	Baltimore	Chicago	Cleveland	Dallas	Detroit	Houston	Los Angeles	Minneapolis-St. Paul	New York	Philadelphia	Pittsburgh	St. Louis	San Francisco
Baltimore	0	0	2	28	0	35	1	7	0	0	14	36	0
	0	*0*	*0*	*40*	*0*	*40*	*0*	*0*	*0*	*0*	*0*	*40*	*0*
Chicago	0	0	17	1	17	0	31	9	17	0	0	0	15
	0	*0*	*40*	*0*	*40*	*0*	*40*	*40*	*0*	*0*	*0*	*0*	*40*
Cleveland	2	17	0	3	35	1	16	25	1	0	1	3	1
	0	*40*	*0*	*0*	*40*	*0*	*40*	*40*	*0*	*0*	*0*	*0*	*40*
Dallas	28	1	3	2	2	27	1	11	0	0	6	24	0
	40	*0*	*0*	*0*	*0*	*40*	*0*	*0*	*0*	*0*	*0*	*40*	*0*
Detroit	0	17	35	2	0	0	14	25	3	1	0	1	3
	0	*40*	*40*	*0*	*0*	*0*	*40*	*40*	*0*	*0*	*0*	*0*	*40*
Houston	35	0	1	27	0	0	0	4	0	0	14	37	0
	40	*0*	*0*	*40*	*0*	*0*	*0*	*0*	*0*	*0*	*0*	*40*	*0*
Los Angeles	1	31	16	1	14	0	1	9	17	0	0	1	15
	0	*40*	*40*	*0*	*40*	*0*	*0*	*40*	*0*	*0*	*0*	*0*	*40*
Minneapolis-	7	9	25	11	25	4	9	1	1	1	6	7	1
St. Paul	*0*	*40*	*40*	*0*	*40*	*0*	*40*	*0*	*0*	*0*	*0*	*0*	*40*
New York	0	17	1	0	3	0	17	1	0	2	0	0	38
	0	*0*	*0*	*0*	*0*	*0*	*0*	*0*	*40*	*0*	*0*	*0*	*0*
Philadelphia	0	0	0	0	1	0	0	1	2	37	0	0	2
	0	*0*	*0*	*0*	*0*	*0*	*0*	*0*	*0*	*40*	*0*	*0*	*0*
Pittsburgh	14	0	1	6	0	14	0	6	0	0	22	14	0
	0	*0*	*0*	*0*	*0*	*0*	*0*	*0*	*0*	*0*	*40*	*0*	*0*
St. Louis	36	0	3	24	1	37	1	7	0	0	14	0	0
	40	*0*	*0*	*40*	*0*	*40*	*0*	*0*	*0*	*0*	*0*	*0*	*0*
San Francisco	0	15	1	0	3	0	15	1	38	2	0	0	1
	0	*40*	*40*	*0*	*40*	*0*	*40*	*40*	*0*	*0*	*0*	*0*	*0*

V. CONCLUSIONS

The results of this study indicate that the appearance of a face can vary markedly depending on the assignment of variables to facial features. Further, it was demonstrated that the ability of people to replicate a statistically derived typology solely on the basis of perceived similarities among faces is also a function of the assignment of variables to facial features. This latter finding suggests that people perceive differences among faces on the basis of certain special facial features which they deem more important than others. This is in direct contrast to perceived differences on the basis of some composite effect. If perceived differences among faces was a function of some composite, then, there wouldn't be any importance associated with the assignment of variables to facial features. It could be that in order to create a meaningful composite, many variables have to be utilized. The results of this study may have been due to the fact that only seven variables were employed. In any event, this is a problem area requiring additional research.

There are several potentially important uses of faces in regional analysis. First, they can be used for multidimensional data representation. Conventional graphical techniques are limited to three dimensions. Second, as was demonstrated in this study, faces can be utilized in regional classification studies. An analyst can obtain an immediate profile of a region from an examination of a single face. In addition, a comparison of several faces makes it possible to detect regional patterns.

Another potentially important use of faces in regional analysis would be for longitudinal analyses. For example, with the increasing attention being given to an assessment of changes in the quality of life of areas, it is possible through the use of faces to have pictorial quality of life indicators for various regions. These pictorial indicators can then be monitored over time. Finally, it may be possible to detect sequential changes temporarily in facial features that might suggest a formal structural process. While certain mathematical techniques can be employed for this latter function, it might be that the eye may be a better "filter" in detecting spatially or temporally distributed patterns in the data.[5]

[5]Ronald M. Pickett and Benjamin W. White, "Constructing Data Pictures," Proceedings of the 7th National Symposium of the Society for Information Display, October, 1976.

THE SYRACUSE PERSON PERCEPTION TEST: A MEASURE OF RESPONSIVENESS TO FACIAL AND VERBAL CUES

Wayne W. Crouch
Bette Brindle

School of Information Studies
Syracuse University
Syracuse, New York

Jerry K. Frye[1]

Department of Speech Communication
SUNY at Buffalo
Buffalo, New York

I. INTRODUCTION

The analysis of multivariate data is being attempted by a number of researchers by using graphical representations of the data. Much of this work has used the cartoon-like face developed by Chernoff (1973) which attempts to take advantage of man's sensitivity to changes in facial expressions; each variable is represented as a facial feature that can vary in some systematic way. Thus, the face as a whole represents the values of the multiple variables simultaneously.

This paper describes the development of a test to measure individuals' responsiveness to facial and verbal cues. The graphical representation of data is only useful to the extent that analysts can see something useful in the representations. The test which is described here is the first step in an attempt to better understand responsiveness to a variety of cues. In this introductory section we present our assumptions

---

[1] Present address: Department of Communication, University of Minnesota, 465 A.B. Anderson Hall, Duluth, Minnesota 55812.

ISBN 0-12-734750-X

in conducting this work and review some relevant work of other researchers. In the second section we describe the test, its development, and the rationale for some of the decisions we made. Finally, we discuss the results which are now available from administering the test to subjects.

A. Communication Cues

The underlying assumptions of this work are that communication involves the processing of cues by individual communicators and that research will be facilitated by our being able to reliably measure how an individual processes these cues. Historically, verbal cues have long been studied and are universally recognized as an important aspect (if not the totality) of human communication. More recently a variety of nonverbal cues have also been identified as important.

Among the numberous types of nonverbal cues identified, facial cues have been the subject of speculation and study to a greater extent than any others. Attention to them dates back at least to Aristotle (Harrison 1973); Darwin's work (1872) is often referred to; and facial cues have been studied vigorously by scientists in the period from about 1914-1940 and from 1960 to date (Ekman, Friesen, and Ellsworth, 1972). During the intervening twenty years, the face seemed to be in "scientific disrepute." Ekman, Friesen, and Ellsworth put it this way:

> Does the face provide accurate information about emotion? Are the facial behaviors related to emotion innate or learned? For a time there was considerable argument in the literatures, but a pessimistic view became dominant. At the least, the results appeared contradictory, and certain advocates--most notable Hunt, Landis, and Sherman--argued that the face was a poor source of information about emotion. There was no accuracy, either as judged by observers or through direct measurement of the face. What little agreement about emotion could be achieved depended more on knowledge of the eliciting circumstance than on observation of the face. And no evidence for innate elements could be found (pp. 7-8).

Books by Ekman and Firesen (1975), Ekman, Friesen, and Ellsworth (1972), and Izard (1971) attest to the resurgence of research on the face.

Optimism about the fruitfulness of studying the face is growing. Ekman (1971), for instance, reports:

> Our findings, supported by those of others, now provide the basis for settling the old dispute as to whether facial expressions are completely specific to each culture or totally universal. Our neurocultural theory maintains there are both universal and culture-specific expressions. The evidence now proves the existence of universal facial expressions.... And it is not simply the recognition of emotion that is universal, but the expression of emotion as well (Ekman, 1971, p. 278).

In addition, Mehrabian (1972) finds that the face is influential in determining "inferred attitudes." From studies (Mehrabian and Weiner, 1967; Mehrabian and Ferris, 1967) of facial, verbal, and vocal cues he derives a regression equation to predict the responses to combinations of all three:

> . . the studies suggest that the combined effect of simultaneous verbal, vocal and facial attitude communications is a weighted sum of their independent effects as follows:
>
> $$A_{total} = 0.07\ A_{verbal} + 0.38\ A_{vocal} + 0.55\ A_{facial}$$
>
> where all four attitude variables are measured on the same scale . . . (Mehrabian, 1972, p. 108).

The contribution of the face dominates his equation, although vocal and verbal contributions are also significant. Levitt (1964) found a similar relationship between vocal and facial cues. And Argyle, *et al*. (1970), likewise, found that nonverbal cues (including facial, postural,and vocal) "had 4.3 times the effect of verbal cues on shift of ratings, and accounted for 10.3 times as much variance."

In addition, Argyle, *et al*., found that females were more responsive to nonverbal cues whereas males were more responsive to verbal cues. Crouch (1976), using a measure of facial versus verbal responsiveness, found a similar sex difference and a relationship between dominant direction and conjugate lateral eye movement and responsiveness.

Thus, the importance of the face has again come under intensive study, and the results suggest that it is an important source of cues upon which reliable inferences about the individual can be based. Facial and verbal cues, then, emerge as promising for study, both in terms of the amount of information accumulated about them and their primacy as sources of cues for making inferences about others.

B. Facial and Verbal Cues

Examination of what is known about facial and verbal cues at present suggests that they may differ considerably in characteristics that are important for efficient information processing. Reusch and Kees (1956) suggested that the distinction between analogic and digital data and data processing schemes was applicable to nonverbal communication and that there was a striking difference between verbal and nonverbal cues.

Linguistic cues seem best represented as digital and arbitrary. Linguists have successfully isolated discrete units (phonemes) upon which spoken language is built, and written language is, of course, based on the alphabet. An iconic relationship between the units and the referents of combinations of units is minimal, if existing at all.

However, a digital micro-structure does not insure that all uses of language require only a digital ability. Actually using language may require digital ability, but knowing what to use it for may be influenced by analogic abilities or digital abilities, or both (Hildum, 1967, p.43).

Facial expressions seem to be continuously related to their referents and consist of a spatial arrangement of features to be processed. Analogic processing using a simultaneous method would seem most appropriate. It is by no means clear, however, that these cues are in fact processed analogically. Birdwhistell (1970, pp. 75-77) assumes all body movements are coded similarly to language. But Ekman and Friesen (1969) suggest that facial affects may be either iconic or arbitrary:

> The coding of facial affect displays is not at all obvious. Both Darwin's explanation of the evolution of such displays, and our account of how certain displays may naturally develop in the course of each person's life,

would suggest that some affect displays are either intrinsically coded or iconic. This may be so only for some affects; if we accept Darwin's principle of anti-thesis as the explanation of the happiness display, then it would be arbitrarily coded.

The channels through which linguistic and facial cues are perceived would seem to be inherently better suited to one of the modes. Paivio (1971, p. 33), for instance, argues that vision employs a parallel processing mechanism whereas hearing employs a sequential processing mechanism.

Thus, in understanding how people respond to cues, the goal is to be able to separate perceptual from inference processes and to specify whether digital or analogic techniques are being used. Also, efficiency may not only be a function of characteristics of the cues, but also of the channels through which the cues are presented. Linguistic cues, for instance, can be transmitted through the visual channel as when reading a transcript of what someone said.

The work described in this paper focuses on facial and verbal cues, both presented visually. Cartoon faces and printed statements are used to measure responsiveness to the cues. In a paper-and-pencil test that can be administered to groups, subjects are asked to give one response to each of several combinations of faces and statements. The response is a function of both the perceptual and inference processes.

C. Measures of Responsiveness to Facial and Verbal Cues

To measure differences between individuals, subjects are given stimuli consisting of linguistic cues, facial cues, or both. And the channel through which the cues are presented is controlled. Such procedures are common in much person perception research (Tagiuri, 1969). A Number of these are described in detail in a previous paper (Crouch, 1974). Briefly, Harrison (1964) studied facial cues alone. He used cartoon type drawings (pictics). Shapiro (1968) studied linguistic and facial cues together. He used the visual channel for both linguistic cues (printed statements) and facial cues (some of Harrison's pictic faces). Vande Creek (1972) adapted Shapiro's general method. He used the auditory channel for linguistic cues (spoken words via videotape). Argyle _et al._, (1970) also used a videotape presentation, but vocal and postural cues were combined with facial cues for study. Shannon (1970) studied facial cues alone. She used the visual channel (photographs of faces). Rosenthal _et al_. (1974) used a motion picture presentation that included the face, the body, randomized spliced voice, and content-filtered voice.

The test describe in this paper is similar to the work of Shapiro, Vande Creek, and Argyle, _et al._, in that subjects are asked to give single responses to combinations of facial and verbal cues. It uses Argyle's design which has a number of advantages over the designs used by Shapiro and Vande Creek. (See Crouch, 1974, for a more detailed discussion). The test itself and reliability and validity data are described in the remainder of the paper.

II. DEVELOPMENT OF THE TEST

As the introductory discussion suggests, verbal and facial cues were chosen from the numerous types of communication cues used in human communication. In addition, a particular mode of presentation was selected. Verbal cues are presented as typewritten statements; facial cues are presented as schematic line-drawings of faces. (See Figure 1.)

For the presentation of verbal cues, audio recordings were considered We could have used spontaneous utterances, acted utterances, or narrated utterances. In the former two cases paralinguistic cues such as tone of voice and inflection would also have been present. Thus, narrated utterances or printed statements seemed the best possibilities. We are interested in comparing responses to the two different modes, but chose printed statements for the initial test development since they used the

Faces

Statements

(1) I'm afraid I'll go to Hell when I think that kind of stuff.

(2) Oh, lunch was about as usual.

(3) I had a good time skiing last weekend.

(4) My father has never thought I'd be good at anything.

(5) I think I should take more courses next term.

(6) I guess I just enjoy myself on days like this - had a real good ride coming down.

A Sample Combination

I guess I just enjoy myself on days like this - had a real good ride coming down.

FIGURE 1. The faces and statements used in The Syracuse Person Perception Test.

visual channel, as do all modes that we considered for presenting facial cues. Thus, our initial test uses the same channel for presentation of both types of cues.

Facial cues could be presented by a "live" actor, or by various methods of recording that differ as to motion versus still and color versus black and white. The recordings could be of spontaneous or acted cues. Drawings can also be used. Similarly, they could be presented with motion or as stills and in color or black and white. In addition, the amount of detail in the drawing can vary. We were interested in comparing responses to the different modes, but we chose a simple line-drawing presentation in black and white for this initial test development. Specifically, we used the faces Harrison (1964) developed. His research shows that individuals can judge emotions from this type of presentation of facial cues.

Following the work of Shapiro (1968), Vande Creek (1972), Argyle *et al.* (1970), and Rosenthal *et al.* (1974), we wanted to measure responsiveness to both types of cues when they were presented simultaneously. If several presentations of verbal cues are combined with several presentations of facial cues into all possible combinations, it is possible to determine the independent effect of each type of cue on the subjects' responses to the combinations. Doing so involves an analysis of variance technique which is described in detail in a previous paper (Crouch, 1974). The technique will be described below for this particular test.

Individuals make a variety of responses to communicative cues. We chose to focus on the judgment of how pleasant the portrayed person was feeling. The judgment of emotion has been widely studied (Ekman and Friesen, 1969) and a number of studies have found "pleasantness" to be an emotion which is, in fact, meaningful to respondents and judged by them (Ekman *et al.*, 1972).

A. Method

Sixty statements and 60 faces[2] were presented separately to a group of 81 subjects.[3] The subjects were instructed as follows:

> The individual quoted[4] in the booklet given to you is a young man of college age taking part in psychological counseling interviews. Rate how pleasant he is feeling in each instance with a score from 1 to 9, giving a rating of 9 when he is feeling extremely pleasant, a 1 when he is feeling extremely unpleasant, and a 5 as a neutral score.

In addition, the following scale was displayed:

unpleasant 1 2 3 4 5 6 7 8 9 pleasant

Two faces and two statements were chosen from each of three points on the scale (1.7, 5.0, and 7.3, $\pm$ .2 for each). All possible combinations (36) of these faces and statements were then formed, and the combinations were presented to another group of subjects (N=41).[5] The order of presentation of the combinations was determined randomly (within certain constraints[6]) for each subject. Figure 1 shows the faces and statements used and a sample combination of a statement and a face. The subjects were instructed as follows:

> Assume that the person on the following pages is a young man of college age taking part in psychological counseling interviews. Rate how pleasant he is feeling in each instance with a score from 1-9, giving a rating of 9 when he is feeling extremely pleasant, and 1 when he is feeling extremely unpleasant, and a 5 as a neutral score.

The scale shown above was also displayed.

---

[2]The 60 statements consisted of 46 from Shapiro (1968) and 14 developed by the first author. The 60 faces were from Harrison (1964).

[3]These subjects were students in undergraduate and graduate courses at a large midwestern university; ages 18 to 56 years with a median of 22.5 years; 38 male, 43 female.

[4]The word "quoted" was used in conjunction with the statements. For the faces the word "portrayed" was used.

[5]These subjects were students in graduate courses in the School of Information Studies,Syracuse University. All were female; their ages were not obtained.

[6]The pages of each test booklet were ordered randomly with the following constraints: for the 36- and 18-item tests neither a statement nor a face could reappear until the third following page, and for the 9-item tests used in study 5 neither a statement nor a face could appear until the second following page.

Mean responses to the 36 combinations are shown in Table 1. Whereas the least pleasant statements alone were rated 2.7, the average rating for combinations including those statements was 3.9 (See Table 2). Similarly, the most pleasant statements alone were rated 7.3, but the average in combination was 5.7. The least pleasant faces alone were also rated 2.7, but the average rating for combinations including those faces was 3.0. And similarly, the most pleasant faces were rated 7.3, but the average of the combinations was 6.4. The greater shrinkage of responses to the statements is consistent with previous finding by Argyle et al., (1970) and Mehrabian (1972).

Each quadrant in Table 1 includes the responses to 9 combinations. The variance of those 9 responses can be partitioned into three components: 1) the variance of the row marginals, 2) the variance of the

		STATEMENTS						
		1 Unpleasant	2 Neutral	3 Pleasant	4 Unpleasant	5 Neutral	6 Pleasant	
FACES	1 Unpleasant	2.5	3.0	4.2	2.3	3.0	4.0	3.2
	2 Neutral	4.1	4.8	5.2	3.5	4.6	5.8	4.7
	3 Pleasant	5.0	6.1	7.6	5.8	6.1	7.5	6.3
	4 Unpleasant	2.1	2.8	3.9	2.0	2.6	3.9	2.9
	5 Neutral	4.2	4.9	5.9	4.3	5.0	5.8	5.0
	6 Pleasant	6.0	6.0	7.5	5.8	6.1	7.4	6.5
		4.0	4.6	5.7	3.9	4.6	5.8	4.8

Table 1: Mean Responses of 41 Female Subjects to Combinations of Statements and Faces.

	When Presented Alone	When presented in Combination
Statements		
1 and 4	2.7	3.9
2 and 5	5.0	4.6
3 and 6	7.3	5.7
Faces		
1 and 4	2.7	3.0
2 and 5	5.0	4.8
3 and 6	7.3	6.4

Table 2: Mean Rating (±.2) of Faces and Statements

column marginals, and 3) the remaining variance, if any, which is due to interaction between the columns and rows. The row marginals are the average responses across all combinations which include the unpleasant face (Row 1), the neutral face (Row 2), and the pleasant face (Row 3). The column marginals are the average responses across all combinations which include the unpleasant statement (Column 1), the neutral statement (Column 2), and the pleasant statement (Column 3). See figure 2.

We use V_s as a measure of a subject's responsiveness to the statements and V_f as a measure of a subject's responsiveness to the faces. We also define a score called Diff which is $V_f - V_s$. Thus, when the subject is presented with all 36 combinations he gets 4 V_f scores, 4 V_s scores, and 4 Diff scores which are averaged to provide a single value for each.

Each quadrant in Table 1 includes the responses to 9 combinations. We wanted to choose two quadrants that did not share any statements or faces in order to have an 18-item test for further use. Such a test has two independent 9-item tests within it. We could choose between 1) the upper left and lower right quadrants and 2) the lower left and upper right quadrants. In addition, we could interchange the 2 faces or the 2

Statements

FACES		1	2	3	
	1	R_{11}	R_{21}	R_{31}	F_1
	2	R_{12}	R_{22}	R_{32}	F_2
	3	R_{13}	R_{23}	R_{33}	F_3
		S_1	S_2	S_3	

V_{total} = Variance of $[R_{11}, R_{12}, R_{13}, R_{21}, R_{22}, R_{23}, R_{31}, R_{32}, R_{33}]$

V_s = Variance of $[S_1, S_2, S_3]$

V_f = Variance of $[F_1, F_2, F_3]$

$V_{interaction} = V_{total} - V_s - V_f$

or $V_{total} = V_s + V_f + V_{interaction}$

where R_{11} = response to the combination of statement 1 and face 1

R_{12} = response to the combination of statement 1 and face 2

.
.
.

R_{33} = response to combination of statement 3 and face 3

S_1 = the average of $[R_{11}, R_{12}, R_{13}]$

.
.
.

F_3 = the average of $[R_{13}, R_{23}, R_{33}]$

FIGURE 2. Partitioning of the Variance of a Subject's Responses

statements which alone were rated at the three points on the pleasantness scale[7]. Our goal was to obtain two independent 9-item tests within which the pattern of responses was as close as possible to the pattern shown in Figure 3 and the correlations of 1) V_{s1} with V_{s2}, 2) V_{f1}, with V_{f2}, and 3) $Diff_1$ with $Diff_2$ were as high as possible.

$V_{s1} = V_s$ in one quadrant

$V_{s2} = V_s$ in the other quadrant

$V_{f1} = V_f$ in one quadrant

$V_{f2} = V_f$ in the other quadrant

$Diff_1 = Diff$ in one quadrant

$Diff_2 = Diff$ in the other quadrant

$Diff = V_f - V_s$

Each of these correlations is a type of split-half reliability measure.

The lower left and upper right quadrants were chosen. Figure 4 shows the pattern of mean responses for the combinations in those quadrants. The pattern is a reasonable approximation of the ideal shown in Figure 3.

Thus, we have an 18-item test which provides two measures of V_f, V_s and Diff. The two measures can be compared for an internal reliability check, and they are averaged to provide each individual's scores on the test as a whole.

In the next section we discuss the results obtained from testing several groups of subjects.

---

[7] This is accomplished by switching rows 1 and 4, or 2 and 5, or 3 and 6 or columns 1 and 4, or 2 and 5, or 3 and 6 in Table 1. We made such switches a number of times before arriving at the pattern shown in Table 1. However, the amount of improvement due to such switching was not great.

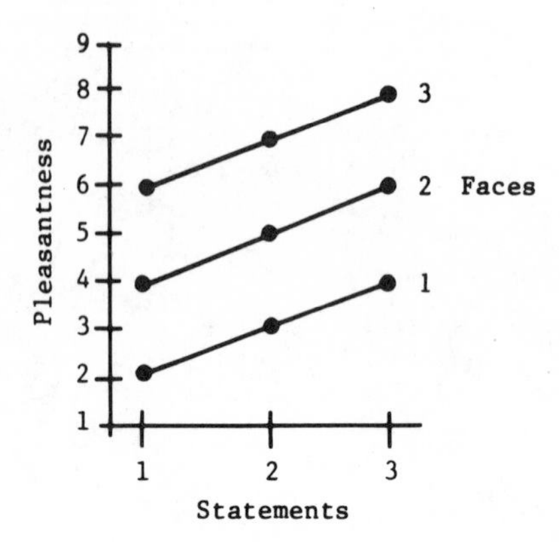

FIGURE 3: Ideal Pattern for Mean Responses to Combinations of Statements and Faces.

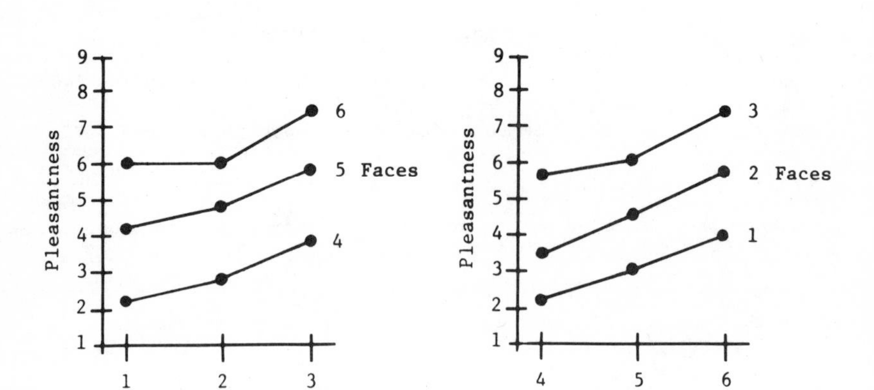

FIGURE 4: Mean Responses to Combinations Chosen for the 18-item Test.

III. TEST RESULTS: SCORES OBTAINED, RELIABILITY, AND VALIDITY

The test has been used in five studies, four of which included female subjects only. In the first two studies the 18 selected combinations were embedded in a test using 36 items, and it was on the basis of these studies that the 18 items were selected. In studies 3 and 4 an 18-item test was given in conjunction with other measures. In study 5 the two 9-item tests were administered separately four weeks apart to the same subjects.[8]

A. Scores Obtained

Two of the scores of interest (V_f and V_s) are variances of three numbers, and each number must fall between 1 and 9 inclusive. Thus, the maximum value of the variance is 42 8/9 (for scores of 1, 1, 9 or 1, 9, 9). Since the combinations presumably fall at regular intervals between some values more central than 1 and 9, a more realistic maximum might be the variance of the numbers 1, 5, and 9. That variance is 10 2/3. Since Diff is $V_f - V_s$, its range would be -10 2/3 to +10 2/3.

Combining studies 3, 4 and 5, scores with the characteristics shown in Table 3 were obtained for 194 subjects. The mean of 3.0 for V_f and 1.4 for V_s indicates that subjects are more responsive to the faces. This is consistent with the previous findings of judgements of the statements and faces alone. (See Table 2). We had hoped that most of the variance of a subject's responses would be account for by V_f and V_s. The remaining variance was, in fact, found to be relatively small for most subjects (mean=0.7, maximum=4.3). This would indicate that the subjects' responses

---

[8]The subjects in Studies 3 (N=29) and 5 (N=64) were all students in graduate courses in the School of Information Studies, Syracuse University. All were female; their ages were not obtained. The subjects in Study 4 (N=101) were high school students in a suburban Syracuse school district. Both males and females were included; their ages were not obtained.

	Mininum	Mean	Maximum	Standard Deviation
V_f	0.0	3.0	10.0	2.2
V_s	0.0	1.4	10.7	1.9
Diff	-10.7	1.6	9.6	3.6

TABLE 3: Range of Scores (N=194).

to these combinations can be well represented as weighted averages of their responses to the statements and faces separately.

The relationships between the scores can be seen in the correlations among them which were obtained in Study 4 (N=101). See Table 4. Subjects who have high facial responsiveness scores tend to have low verbal responsiveness scores. The low correlations of $V_{interaction}$ with other scores supports our practice of ignoring it as a "meaningful" score. We view it as random error.

	V_s	V_f	V_{total}	$V_{interaction}$	V_{diff}
V_s	-	-.45*	.31	.12**	.79
V_f		-	.64	-.14**	-.90
V_{total}			-	.32	-.29
$V_{interaction}$				-	.26**
V_{diff}					-

*significant, =.05, two-alternative test
**not significant, =.05, two-alternative test
NOTE: The correlation coefficients without asterisks involve part-whole correlations.

TABLE 4: Relationships Between the Responsiveness Measures.

B. Reliability

In studies 1, 2, 3, and 4 the correlations between V_{s1} and V_{s2} varied from .67 to .90. The correlations between V_{f1} and V_{f2} varied from .44 to .75. And the correlations between $Diff_1$ and $Diff_2$ varied from .64 to .88. In study 5 the correlations over a four week period were .35 for V_{s1} and V_{s2}, .43 for V_{f1} with V_{f2}, and .45 for $Diff_1$ with $Diff_2$. All of these correlations plus ones for $V_{total\ 1}$ with $V_{total\ 2}$ and $V_{interaction\ 1}$ with $V_{interaction\ 2}$ are listed in Table 5.

Pearson Product Moment Correlations

	V_{s1} with V_{s2}	V_{f1} with V_{f2}	$Diff_1$ with $Diff_2$	V_{total1} with V_{total2}	$V_{interaction1}$ with $V_{interaction2}$
Internal Reliability Measures					
Study 1 N=19	.85	.75	.88	.68	.63
Study 2 N=22	.90	.44	.68	.45	.40
Study 3 N=36	.81	.67	.79	.67	.03
Study 4 N=101	.67	.59	.64	.62	.38
Test-Retest Reliability Measures					
Study 5 N=64	.35	.43	.45	.25	.04

TABLE 5: Internal Reliability Measures and Test-Retest Reliability Measures

C. Validity

We are interested in using this test to explore a number of theoretical positions relevant to human information processing and communication. To date, we have tested subjects on the Eysench Personality Inventory dimensions of extraversion and neuroticism, on the Concealed Figure Test which was developed as a measure of field-independence, and on FLAGS which is a test of visual-spatial abilities. Correlations between these scores and Syracuse Person Perception Test Scores are shown in Table 6 for Study 3. In that study we tested 29 female subjects. With a sample that small the correlations obtained are not statistically significant.

Pearson Product Moment Correlations

	V_s	V_f	Diff	Neu-rot-icism	Extra-version	Field Inde-pend-ence	Visual-Spatial Ability ("Flags")
V_s	-	-.46	-.80	.00	-.33	-.24	+.02
V_f		-	+.82	+.43	+.29	+.05	+.08
Diff			-	+.22	+.26	+.14	+.07
Neuroticism				-	-.03	-.06	+.02
Extraversion					-	-.18	+.04
Field Independence						-	+.44
Visual-Spatial Ability (The "Flags" Test)							-

TABLE 6: Correlations Between Syracuse Person Perception Test Scores and Several Other Variables from Study 4 (N=29 female subjects)

IV. IMPLICATIONS

A method of measuring responsiveness to different types of communication cues has been described for the case of verbal and facial cues. Initial evaluation of the test suggests that subjects respond to the test in an orderly fashion. And it may be sufficiently reliable to be used in exploring issues related to information processing and communication. In addition, the method offers the opportunity to develop similar tests with other cues presented through different channels and at different levels of iconicity. Each of these issues will be discussed briefly below.

The graphical analysis of multivariate data using Chernoff faces relies on an analyst's ability to recognize changes in facial expressions. This test and other work on "responsiveness" to various types of cues may be helpful in designing presentations of data and in choosing and training data analysts.

Another information processing issue that this particular test may be relevant to is the relationship between cognitive style and communication behavior. Crouch (1976) found a relationship between conjugate lateral eye movements and facial/verbal responsiveness using Shapiro's (1968) test. Conjugate lateral eye movements are thought to be indicators of right or left cerebral dominance which may specify cognitive styles or modes of information processing. That study should be replicated with this instrument, and additional exploration of theoretical and empirical relationships among 1) conjugate lateral eye movements, 2) other measures of hemisphere dominance, 3) verbal responsiveness, 4) facial responsiveness, and 5) other variables should be explored.

Field independence/dependence and dogmatism are other cognitive style variables that may relate to responsiveness to facial and verbal cues. Our study showed a moderate negative (though not statistically significant) relationship between a measure of field independence and verbal responsiveness. Other variables we have looked at are visual-spatial ability (as measured by the "FLAGS" test), neuroticism and extraversion.

The test would also seem to have relevance to at least two person-to-person communication concerns. First, compatibility may in part be a function of individuals responding to the cues that best represent their partner's feelings. Some individuals may make their feelings known primarily through facial expressions and thus require a partner who is responsive to faces. Others may make their feelings known verbally and thus require a partner who is responsive to statements. Such compatibilities might be important between marriage partners, supervisor and subordinate, and student and teacher. Secondly, from a somewhat different point of view some occupations may require one or both types of responsiveness. Counselors and social workers, for instance, may be more successful 1) if they are responsive to facial cues, or alternatively, 2) if they are responsive to both facial and verbal cues and can choose between them with different people in different situations.

In terms of further development of this and similar tests, there are two primary issues to be considered. First, subjects may respond differently depending upon how cues are presented. And second, there are at least four types of cues that should be studied.

There are three concerns relevant to how the cues are presented --iconicity, channels, and ancillary conditions. Iconicity refers to the similarity of a symbol and the thing it represents. Thus, a videotape record is more iconic to the actual presence of another person than a still photograph which in turn is more iconic than a cartoon. The effect of iconicity has been of concern to scholars studying the judgment of emotion in the human face (i.e., Ekman, Friesen, and Ellsworth, 1972, pp. 49-51). It is an open question as to whether individuals respond the same to representations of the face at the various levels of iconicity.

The channel through which the verbal cues are presented may also affect responsiveness. Paivio (1971), among others, argues that audio and visual information processing systems are significantly different. Verbal cues presented audibly may elicit different responses than the same cues presented in printed form.

Finally, there are a whole host of ancillary conditions that may affect an individual when he responds to cues. These conditions may lead to more or less anxiety in the individual or more or less concern with being accurate. Such conditions may well affect his relative responsiveness to various types of cues.

There are numerous types of cues that could be studied. Verbal and facial cues have been of central concern in much communication research. Postural cues (Rosenberg and Langer, 1965; Mortensen, 1972, pp.225-226) and paralinguistic cues (Mortensen, 1972, pp. 228-229) are also known to convey information. Thus, it would seem that at least these four types should be of immediate concern.

Communication researchers have long considered the verbal or nonverbal nature of cues to be significant. An important question about responsiveness is whether or not individuals are more or less responsive to all nonverbal cues (i.e. facial, paralinguistic, and postural) as compared with verbal cues. An alternative is that individuals are more or less responsive to cues in one channel. In the latter case verbal and paralinguistic cues (auditory) would pair up against facial and postural cues (visual). The nonverbal and auditory nature of paralinguistic cues makes them central to a test of these two possibilities. Of course, it is also possible that individuals are more or less responsive to each type of cue independent of the verbal/nonverbal and channel distinctions.

Obviously, our comments in this final section are speculative. We suspect, however, that the processing of cues is an important sub-process of communication and that this test can help us better understand human information processing and interpersonal communication.

REFERENCES

Argyle, M., Salter, V., Nicholson, H., Williams, M., and Burgess, P., (1970). "Communication of Inferior and Superior Attitudes by Verbal and Nonverbal Symbols," British Journal of Social and Clinical Psychology, 9:221.

Birdwhistell, Ray L., (1970). "It depends on the Point of View," Kinesis and Context, University of Pennsylvania Press, Philadelphia, pp. 65.

Chernoff, Herman, (1973). "The Uses of Faces to Represent Points in k-Dimensional Space Graphically," Journal of the American Statistical Association, 68:361.

Crouch, Wayne W., (1974). "Individual Differences in Responsiveness to Communicative Cues." A paper presented before the International Communication Association Convention, New Orleans.

Crouch, Wayne W., (1976). "Dominant Direction of Conjugate Lateral Eye Movements and Responsiveness to Facial and Verbal Cues," Perceptual and Motor Skills, 42:167.

Crovitz, Herbert F. and Zener, Karl, (1962). "A Group-Test for Assessing Hand- and Eye-Dominance," American Journal of Psychology, 75:271.

Darwin, C. (1965). The Expression of Emotions in Man and Animals. John Murry, London, 1872. (Republished by University of Chicago Press, 1965).

Ekman, Paul, (1971). "Universals and Cultural Differences in Facial Expressions of Emotion," in James K. Cole (Ed.), Nebraska Symposium on Motivation, University of Nebraska Press, Lincoln, pp. 207.

Ekman, Paul and Friesen, W. V., (1969). "The Repertoire of Nonverbal Behavior: Categories, Origins, Usage, and Coding," Semiotica, 1:49.

Ekman, P. and Friesen, W. V., (1975). Unmasking the Face. Prentice-Hall.

Ekman, P., Friesen, W. V. and Ellsworth, P., (1972). Emotion in the Human Face: Guidelines for Research and An Integration of Findings. Pergamon, New York.

Grumperz, John T., (1968). "The Speech Community," in David L. Sills (Ed.), International Encyclopedia of the Social Sciences, The MacMillian Company and the Free Press, 9:381.

Harrison, Randall P., (1964). Pictic Analysis: Toward a Vocabulary and Syntax for the Pictorial Code, with Research on Facial Communication. Ph.D. Dissertation, Department of Communication, Michigan State University.

Harrison, Randall P., (1973). "Nonverbal Communication," in Ithiel de Sola Pool et al. (Eds.), Handbook of Communication, Rand McNally, pp. 93.

Hildum, Donald C., (1967). Untitled Comment, in Donald C. Hildum (Ed.), Language and Thought, Van Nostrand Reinhold, New York.

Izard, Carroll E., (1971). The Face of Emotion. Appleton-Century-Crofts, New York.

Levitt, Eugene A., (1964). "The Relationship Between Abilities to Express Emotional Meanings Vocally and Facially," in Joel R. Davitz (Ed.), The Communication of Emotional Meaning, McGraw-Hill, pp. 87.

Mehrabian, Albert, (1972). Nonverbal Communication. Aldive-Atherton, Chicago.

Mehrabian, Albert and Ferris, Susan R., (1967). "Inference of Attitudes From Nonverbal Communication in Two Channels," Journal of Consulting Psychology, 31:248.

Mehrabian, Albert and Wiener, Morton, (1967). "Decoding of Inconsistent Communications," Journal of Personality and Social Psychology, 6:109.

Mortensen, C. David, (1972). Communication: The Study of Human Interaction. McGraw-Hill.

Paivio, Allan, (1971). Imagery and Verbal Processes. Holt, Rinehart, and Winston, New York.

Rosenberg, B. G., and Langer, J., (1965). "Study of Postural-Gestural Communication." Journal of Personality and Social Psychology, 2:593.

Rosenthal, Robert, et al., (1974). "Assessing Sensitivity to Nonverbal Communication: The PONS Test," American Psychological Association Division 8 Newsletter, pp. 1.

Ruesch, Jurgen, and Kees, Weldon, (1956). Nonverbal Communication. University of California, Los Angeles.

Shannon, Anna Marian, (1970). Differences Between Depressives and Schizophrenics in the Recognition of Facial Expression of Emotion. Ph.D. Dissertation, University of California, San Francisco. As abstracted in Dissertation Abstracts, 32B:2822.

Shapiro, Jeffrey G., (1968). "Responsivity to Facial and Linguistic Cues," Journal of Communication, 18:11.

Tagiuri, Renato, (1969). "Person Perception," in Gardner Lindzey and Elliot Aronson (Eds.), The Handbook of Social Psychology, Second Edition, 3:1.

Vande Creek, Leon D., (1962). Responses to Incongruent Verbal and Nonverbal Emotional Cues. Ph.D. Dissertation, University of South Dakota.

A STATISTICAL MULTIDIMENSIONAL SCALING METHOD BASED ON THE SPATIAL THEORY OF VOTING

Lawrence Cahoon

U.S. Census Bureau
Suitland, Maryland

Melvin J. Hinich[1]

Department of Economics
Virginia Polytechnic Institute and State University
Blacksburg, Virginia

Peter C. Ordeshook

S.U.P.A.
Carnegie-Mellon University
Pittsburgh, Pennsylvania

A MULTI-DIMENSIONAL STATISTICAL PROCEDURE FOR SPATIAL ANALYSIS

The methodological theory and application discussed in this paper is motivated by the spatial theory of voting. Specifically, we develop a statistical procedure that recovers from cross-sectional survey thermometer score data the spatial configuration of candidates and citizens, as well as the dimensionality of the issue space and the relative salience of these issues.

Briefly, spatial theory seeks to ascertain the policies candidates adopt in an election. The principal assumption of the theory is that candidate strategies and citizen preferences can be represented in an

---

[1]The development of the statistical methods was supported in part by the Office of Naval Research under contract. The methodology used in this paper is discussed in greater detail in a 1975 V.P.I. working paper, "A Multidimensional Statistical Procedure for Spatial Analysis" by the above authors.

ISBN 0-12-734750-X

n-dimensional Euclidean coordinate system with a Euclidean distance metric used to describe citizen utility functions (Davis and Hinich, 1966; Davis, Hinich and Ordeshook, 1970). We cannot review here all of spatial theory's conclusions. Its best known result, however, is that, under a variety of assumptions about citizen preferences and candidate objectives, candidates should adopt policies at or near the electorate's median preference on each dimension.

Using various multi-dimensional scaling methods (eg. [3] and [10]) political scientists are beginning to study spatial models. The most relevant examples are Rusk and Weisberg's analysis of SRC thermometer data, and Rabinowitz's nonmetric algorithm for recovering candidate spatial positions from mass survey data (Rabinowitz, 1974; Rusk and Weisberg, 1972). With the exception of Carroll and Chang, 1970 and Rabinowitz, 1974, multidimensional scaling papers generally avoid the parametric modeling approach advocated by econometricians and most statisticians. Our principal objective here is not simply to develop an alternative algebraic scaling model but rather to formulate a statistical methodology for estimating the parameters of a theoretical model of election competition.

The spatial voting model assumes that there exists a cardinally defined issue space common to all voters and that individual preferences can be represented by the weighted Euclidean distance metric:

$$\|\underset{\sim}{\theta}_j - \underset{\sim}{x}_m\|_A^r = [(\underset{\sim}{\theta}_j - \underset{\sim}{x}_m)'\underset{\sim}{A}(\underset{\sim}{\theta}_j - \underset{\sim}{x}_m)]^{r/2}$$

where $\underset{\sim}{\theta}_j = (\theta_{j1},\ldots,\theta_{jn})'$ denotes the position of candidate j on each of n dimensions, $\underset{\sim}{x}_m = (x_{m1},\ldots,x_{mn})'$ denotes the mth voter's ideal point, and $\underset{\sim}{A} = (a_{ii})$ is a diagonal matrix of positive issues weights. Voters may define the space with a particular sensitivity to positions that are "far" from their ideal points or alternatively they may be sensitive, i.e., perceive differences as substantively meaningful, only if positions are "near" their ideal points. To accomodate the several possibilities, we permit r to vary. If voters are more sensitive to positions that are far from x_m, we let r = 2 (Figure 1a). If the sensitivity is uniform, r = 1 (Figure 1b); and if voters are more cognizant of differences near their ideal points, then we let r = ½ (Figure 1c). Since we do know the appropriate value for r in a given population, it must be estimated.

Suppose now that N voters rate each of p + 1 candidates on a thermometer scale (that varies from 0 to 100), where p > n. Assume that voter m's thermometer score for candidate j is

$$T_{jm} = 100 - \| \underset{\sim}{\theta}_j - \underset{\sim}{x}_m \|_A^r + \varepsilon_{jm},$$

where ε_{jm} is a stochastic zero mean error term.

Our method estimates the dimensionality of the issue space n, each candidate's position in the space $\underset{\sim}{\theta}_j$, each citizen's ideal point $\underset{\sim}{x}_m$, and the issue weights, a_{ii}. In addition, the orientation of the underlying coordinate system with respect to vertical/horizontal axes is estimated when n = 2 and $a_{11} \neq a_{22}$. Our two-dimensional maps are not rotationally invariant as is the case with the other maps produced from the same data base. From now on, let $a_i = a_{ii}$.

We begin with two assumptions: (1) <u>all citizens have the same perception of each candidate</u>; and (2) all citizens weight the dimensions in an identical fashion. Clearly, these two assumptions, implicit in aggregate metric scaling techniques, are restrictive. Without the assumption of some structure, however, estimation is impossible. As a partial resolution of the problem when analyzing the SRC thermometer data, we divide the sample by party identification to check whether there are significant differences in $\underset{\sim}{A}$, in the $\underset{\sim}{\theta}$'s, and the orientation of the axes.

Using only thermometer data to develop a space, there is necessarily an ambiguity between $\underset{\sim}{A}$ and the covariance matrix $\underset{\sim}{\Sigma}$ of the population ideal points. That is, the units of $\underset{\sim}{A}$ are defined in terms of $\underset{\sim}{\Sigma}$ and vise-versa,

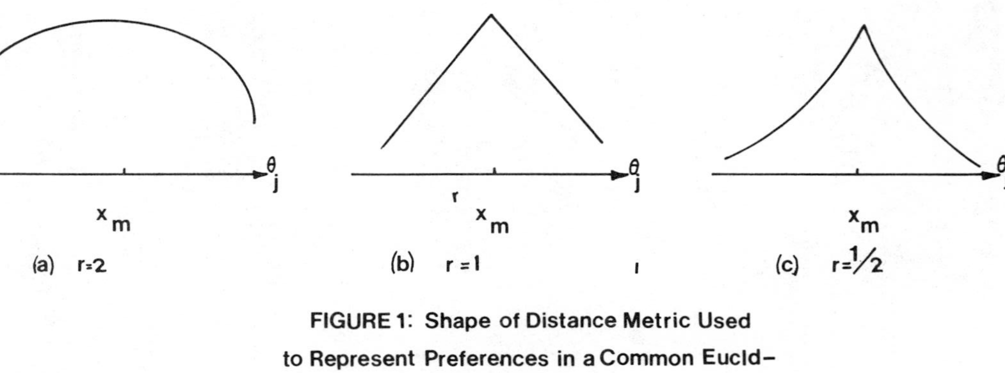

FIGURE 1: Shape of Distance Metric Used to Represent Preferences in a Common Eucld–ean Space

or, more precisely, $\underset{\sim}{A}^{\frac{1}{2}} \Sigma \underset{\sim}{A}^{\frac{1}{2}}$ can be estimated but not $\underset{\sim}{A}$ and $\underset{\sim}{\Sigma}$ separately. Without loss of generality, then, we can set $\underset{\sim}{\Sigma}$ equal to the identity matrix $\underset{\sim}{I}$ and estimate $\underset{\sim}{A}$. This defines $\underset{\sim}{A}$ in terms of standard deviation units. It is not surprising that constraints have to be placed on the space since there are no natural units to the axes.

The method transforms the thermometer scores to a variates which are linear in $\underset{\sim}{\theta}_j$ and $\underset{\sim}{x}_m$. The transformation is simplest when $r = 2$, and is most complex when $r = \frac{1}{2}$. The methodology is described in detail in Cahoon's PhD thesis (1975). Details of the $r = 1$ method is given in Cahoon and Hinich, 1976 and Hinich, 1978. Due to space limitations the statistical details are omitted. It should be emphasized that the estimates of the θ_j's and the a_i are consistent as $N \to \infty$, and are computed using a factor analysis of the covariance matrix of the transformed scores and a set of least-squares fits using mean scores. Although the method is complicated, a Fortran computer program with considerable documentation has been successfully run on three different IBM systems.

In order to proceed with the parameter estimation, the dimensionality of the space must be determined. This is done by studying the eigenvalue pattern of the sample covariance matrix and, more importantly, the eigenvalues of the matrix computed from the transformed data. The problems of selecting the dimensionality of the space and identifying the dimensions in terms of real issues is similar to that encountered using factor analysis. Even when other issue and perceptual data are available, there is no mechanical substitute for scientific judgment and experience.

Once the candidate positions and the other parameters of the space are estimated, the mth voter's ideal point is estimated by regressing the p transformed scores for each m on $\hat{a}_1 \hat{\theta}_{ji}, \ldots, \hat{a}_n \hat{\theta}_{jn}$. As long as $p > n$, a solution can be obtained but the degrees-of-freedom $p - n$ will be small in most applications. Thus, there will be a sizable error in the estimate of a given respondent's ideal point. The estimated ideal points of the respondents should not be trusted, but we can make substantive comments about the average ideal point of a group of respondents if we group the respondents in a meaningful way.

Given the small number of degrees-of-freedom in the ideal point regression, we should reconsider the assumption that all respondents have similar perceptions of all candidates. Although the misperception errors average out in the estimation of the candidate map, there are not enough replications for a given respondent to reduce significantly the error in the ideal point estimate. Since we know that some candidates are better known than others, we use a reduced set of candidates to obtain the ideal points for the 1968 survey--Nixon, Humphrey, Rockefeller, Johnson, Agnew, and Muskie (Robert Kennedy is eliminated since he had been assassinated several months before the survey).

We have not specified how an appropriate value of r is chosen. In applying our method to real data we estimate r simply by trying different values and ascertaining which value yields the closest fit with the internal checks on our assumptions. These checks include: First, all estimated a_i's should be positive. Second, if $\underset{\sim}{x}_m$ is near $\underset{\sim}{\theta}_j$, the reproduced thermometer score should be near 100. Finally, the mean thermometer score of candidate j using the estimated parameters should correlate significantly with j's mean thermometer score. Before we turn to the analysis of artificial and real data, however, in the next section we offer two important caveats to the model.

Valence Dimensions

Suppose that there exists dimensions over which candidate positions vary but voters have identical ideal points. Suppose, for example, that $\underset{\sim}{\theta}_j = (\theta_{1j}, \theta_{2j}, \theta_{3j})$ but that $\underset{\sim}{x}_m = (x_{m1}, x_{m2}, 0)$ for all respondents--i.e., on the third dimension the ideal point of all citizens equals zero. Thus, the variance of the x's on one dimension is zero. The sample covariance matrix fails to reveal the true dimensionality of the space. Moreover our estimates of the mean preference on dimensions 1 and 2, $\bar{x}_1$, and $\bar{x}_2$, contain the effects of the third dimension, and $\theta_{3j}^2 a_3$ cannot be estimated if θ_{3j} is different for each candidate.

For the 1968 survey scores, we hypothesize the existence of a valence issue that conforms to candidate familiarity and assume that $\theta_{j3} = 1$ for the remaining candidates. This adds an intercept to the linear equation which is fitted by least-squares in order to generate estimates of the a_i's and the rotation. If these assumptions are correct, then the remaining issue weights should be estimated properly. We consider, of course, several alternative partitions of candidates and choose the partition that yields positive estimates of the a_i's and a high correlation between average predicted and observed thermometer scores.

Aside from the desirability of estimating distances as opposed to relative positions, one advantage of our procedure is that the estimates of the issue weights a_i provide an internal check for the possible existence of valence issue. Rabinowitz and Rusk and Weisberg do not consider valence issues because their nonmetric method would not indicate the presence of such an issue. Here, however, a negative $\hat{a}_i$ indicates a misspecification in the model underlying the methodology. If the valence issue had not worked with a politically meaningful grouping, the spatial model would have been rejected.

While this procedure is somewhat ad hoc, there are, in fact, sound theoretical reasons for believing that "familiarity" is an issue and that it should receive special treatment in an analysis of election survey data. Our method assumes that citizens possess well-defined perceptions of each candidate's position. In reality, however, citizens are uncertain about positions so that if they are risk averse, respondents should discount thermometer scores by the degree of uncertainty they associate with a candidate. Presumably, however, they perceive some variation among candidates on this criterion. Hence, the perceived riskiness of candidates contaminates the results. But, to the extent that a _qualitative_ assessment of familiarity measures this perception, we can minimize the extent of the contamination.

Artificial Data

In this section we examine the robustness of the method by applying it to artificial data. In the space allowed, we cannot report all test runs; instead we compare one case in which the data satisfy all assumptions to runs in which alternative key assumptions are violated. We begin by generating a set of two-dimensional preferences using a normal N(0,1) random number generator for a sample size of 1000 respondents. Ten candidates are arbitrarily placed in the space, and, letting $r = \frac{1}{2}$ and $a_1/a_2 = .6$ thermometer scores are computed for each respondent. After rescaling these scores so that they fall in the range [0, 100], we then add a normal error term whose standard deviation is 10.95. Each resulting score is then rounded to a member of the set {0, 5, 10, 15, 20, 30, 40, 50, 60, 70, 80, 85, 90, 95, 100}. This procedure eliminates scores less than 0 and greater than 100 and produces data that conform to people's apparent tendency to round their responses.

Applying the method to this data--data that satisfy all assumptions except for rounding--we assume first that the true value of $r = \frac{1}{2}$ is known. We present the results of that run in Figure 2, where dots denote true candidate positions and circles denote recovered positions. To indicate the effects of rounding we add to Figure 2 (and denote by triangles) the positions we recover when thermometer scores are not rounded. Clearly, rounding distorts somewhat our recovery of spatial locations but, overall, the distortion does not appear significant. One measure of its significance is the ratio $| \hat{\theta}_j | / | \theta_j |$ and, in particular, the maximum and minimum observed for this ratio. With respect to the preceding data, all such ratios fall in the range (.88, 1.2)--i.e., we observe at most a 20 percent error. Our estimate of a_1/a_2, moreover, .632, is surprisingly close to its true value .60, while the R^2, between the candidate's average adjusted scores and their estimated values is .9995. Overall, then, the method appears to be quite satisfactory, at least if all assumptions are satisfied and r is known.

Suppose, however, that r is not known. Specifically, in Figures 3, 4, and 5 we contrast recovered and actual candidate positions when r = ½ but is mistakenly assumed to equal 2, 1, and then ½ . Comparing these figures to Figure 2 we observe some deterioration in our estimation, but this deterioration does not appear to be substantial, i.e., the range of $|\hat{\theta}_j| / |\theta_j|$ becomes (.80, 1.34), or a maximum error of 35 percent. The accuracy of our estimates of $\hat{a}_1/\hat{a}_2$ also decline somewhat (equaling .643, .738, and .570 for r = 2, 1, and ½ respectively) although again this decline is not disturbing. More important, however, R^2 increases as well (equaling .935, .994, and .996). Thus, if we use this criterion to select an appropriate value for r, we would in fact choose the correct value. Overall, then, the procedure works well if all assumptions are satisfied but when r must be estimated.

Despite these positive results, no statistical methodology can be wholly insensitive to violations of its assumptions. For an example, consider the assumption that A is diagonal, i.e., that there exists a linear transformation of the axes that renders A diagonal and $\underset{\sim}{\Sigma} = \underset{\sim}{I}$. Suppose instead that with A diagonal,

$$\underset{\sim}{\Sigma} = \begin{bmatrix} 1.0 & \rho \\ \rho & 1.0 \end{bmatrix}$$

where $\rho \neq 0$. Letting $a_2/a_1 = 3.0$, in Figure 6 we contrast the candidate's true and recovered positions for $\rho = .64$. Clearly, correlation in $\underset{\sim}{\Sigma}$ occasions considerable distortion of the recovery (note, though, that a

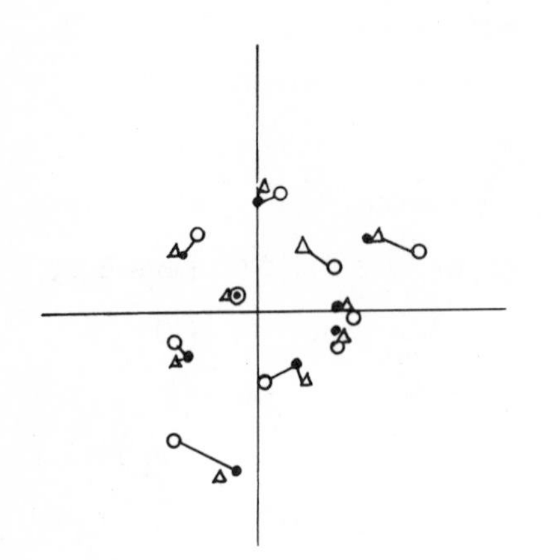

FIGURE 2: r ½ and assumed to be known

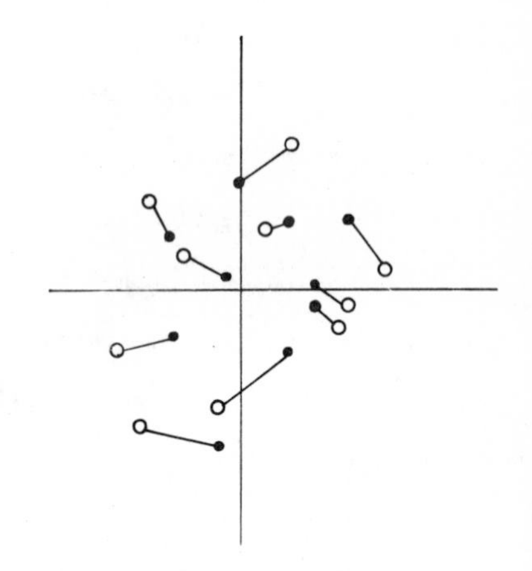

FIGURE 3: r ½ but assumed to be 1

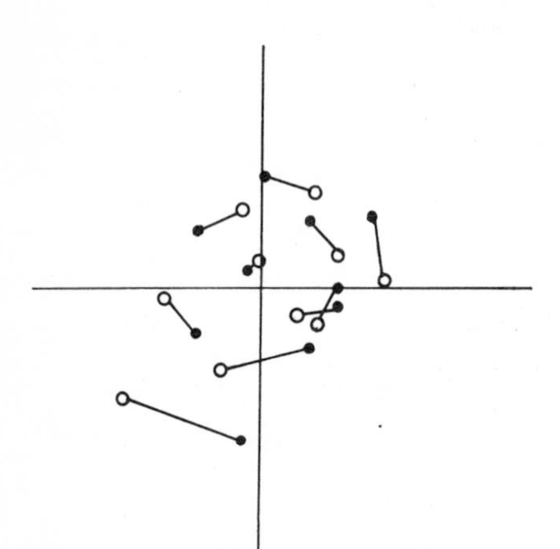

FIGURE 4: r ½ but assumed to be 2

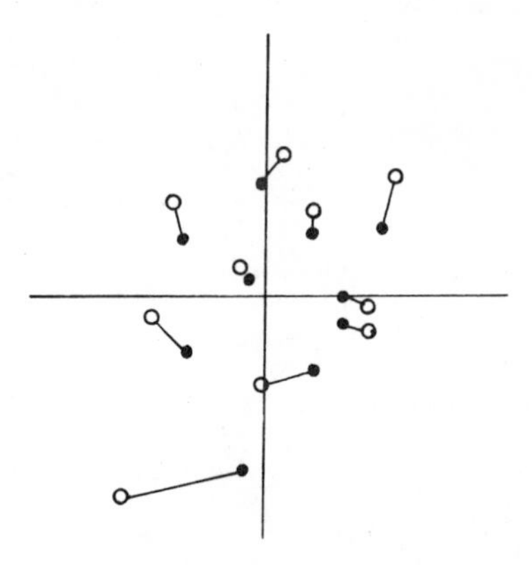

FIGURE 5: r ½ but assumed to be 8

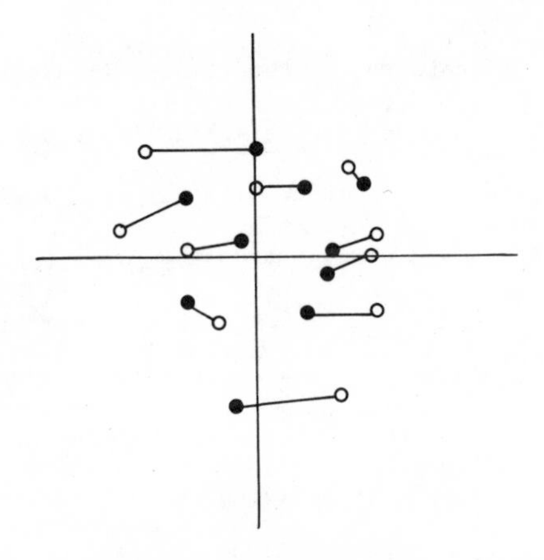

FIGURE 6: Correlation in Σ

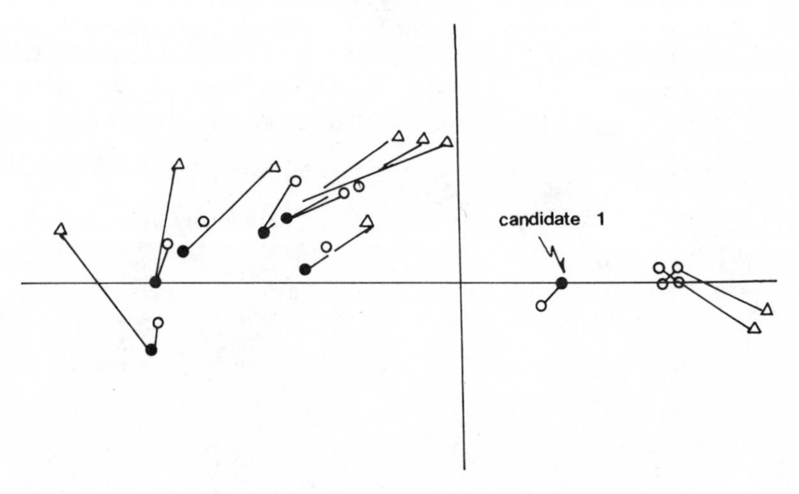

FIGURE 7: Estimated Candidate Positions with Valence Issue - - Corrected (o), Uncorrected (Δ)

substantial amount of the distortion can be attributed to rotational error). Further our estimate of a_2/a_1 fares badly--equaling 7.11 as against the true value of 3.0.

The final artificial data run we report here illustrates the situation we encounter later with valence issues. First, we choose a new two-dimensional configuration of candidates that is not dissimilar from the configuration obtained in the next section. Second, we generate a sample of artificial thermometer response scores as before after letting $a_1/a_2 = 2.33$ and after assuming that there exists a valence issue such that $\theta_{j3} = 1$ for some candidates, $\theta_{j3} = 0$ for others, $x_{m3} = 0$, and $a_3/a_2 = 1.00$. The results now of applying the procedure to this data knowing $r = \frac{1}{2}$ but without a correction for the valence issue yields the negative ratio $\hat{a}_1/\hat{a}_2 = -3.0$ and considerable distortion in our estimates of $\underset{\sim}{\theta}_j$ and the mean. While we cannot use the distortion in $\underset{\sim}{\bar{x}}$ and $\underset{\sim}{\theta}_j$ as a test for valence issues (since we cannot observe the true values of $\underset{\sim}{\bar{x}}$ and $\underset{\sim}{\theta}_j$), a negative estimate for an issue weight is a clear warning. Figure 7 shows, in fact, that even if we overlay candidate 1's recovered and true positions, our recovery of the remaining candidate positions (dots denote actual positions and triangles denote recovered positions) is unsatisfactory.

Suppose, however, that we can properly identify the positions of candidates on the valence issue and rerun the procedure accordingly. Now, $\hat{a}_1/\hat{a}_2$ equals 1.79 while $\hat{a}_3/\hat{a}_2$ equals 1.11 (as against the true values of 2.33 and 1.00). Further, as Figure 7 shows, the recovery of candidate positions (denoted by circles) is improved considerably (without any correction for the origin of the space). We conclude, then, that perhaps even more than off diagonal elements in $\underset{\sim}{\Sigma}$ or misidentification of r, valence issues can render the results of our procedure (and presumably any similar procedure that uses thermometer data) distorted or meaningless.

But, if such an issue exists and if we can independently estimate or guess the positions of the candidates on it, the procedure can be modified appropriately.

Election Data

We know, of course, that no statistical procedure can be applied indiscriminately to data. The appropriateness of a particular procedure--such as linear regression analysis--depends on our research interest and on our willingness to assume that the data conform to the statistical and structural assumptions of the procedure. Thus, beyond examining artificial data, we cannot "test" the methodology developed in the preceding sections in the sense that we test a theory. Instead, we must apply it to data that we assume are appropriate and assess the method's adequacy in more intuitive ways. First, we require that the resultant spatial maps make some sense. That is, if our estimates of the candidates' positions depart significantly from intuitive presuppositions or from the estimates others obtain using different techniques, then we must either formulate an explicit hypothesis about why those techniques are inappropriate or consider the hypothesis that our assumptions do not adequately approximate the data's properties. Second, all estimated a's must be positive.

We begin by noting that the results of a straightforward application of our procedure to the 1968 SRC election survey data is unsatisfactory in several respects. First, while two dimensions are estimated, the a_i's for one or more of these dimensions are negative. Second, much like the results reported elsewhere, George Wallace and Curtis LeMay are located on a separate dimension, far from other candidates. Third, the distances between the $\hat{\underset{\sim}{\theta}}_j$'s and $\underset{\sim}{\bar{x}}$ do not correlate highly with the candidates' average thermometer scores across respondents.

A simple examination of the data, however, suggests immediately why a naive application of the procedure is inappropriate. First, the data seem overly noisy despite our allowances for noise in the model. In particular, far too many respondents assign scores of fifty to all but a few candidates--suggesting that they either possess no well-defined attitudes or that they fail to respond and discriminate carefully. Our resolution of this problem is to drop from the sample all respondents who report a score of fifty for four or more candidates (out of 12). In addition, respondents who did not score all the candidates named were removed since the model assumes that the respondent be politically aware.

The second problem with the data concerns the Wallace and LeMay responses. Consider Table 1, where we report the number of zero scores given each candidate by the filtered sample's respondents.

Clearly, Wallace and LeMay are distinct from all other candidates in this respect. While the respondents giving zero scores to these two running mates number in the hundreds, their closest "competitor" is Reagan with a total of fifty-six zero scores. This suggests that a significant percentage of the response scores for these two candidates do not conform to our model's assumptions. We thus decided to delete the scores for Wallace and LeMay.

A third potential problem concerns the assumption that all respondents share a common $\underset{\sim}{A}$-matrix. We know, of course, that such an assumption is probably false with respect to "actionable issues" such as Vietnam, busing, or inflation. Our procedure, however, seeks to recover underlying dimensions of taste that, in conjunction with socio-economic circumstances among other things, presumably determine a person's preferences on

TABLE 1. Number of Respondents Reporting Zero Thermometer Scores by Candidate, 1968

	Full Sample	Democrats	Independents	Republicans
Wallace	265 (43%)	142	73	97
LeMay	181	107	46	59
Reagan	56	44	18	7
McCarthy	44	19	14	21
Johnson	43	11	16	21
Romney	42	22	16	13
Agnew	41	35	12	3
Humphrey	37	7	17	24
Rockefeller	34	16	11	14
Kennedy	32	12	15	16
Muskie	23	9	8	12
Nixon	16	15	3	0
N	620	317	206	244

Note: Due to the overlap in these groups the number of Democrats, Independents, and Republicans exceed 620.

actionable issues. There is little if any evidence to suggest that the relative importance of these dimensions varies systematically across the electorate (if only because methods for discussing such variations do not exist). Nevertheless, as a partial control for variations in $\underset{\sim}{A}$, we divide the sample into three subsamples--Democratic and Republican party identifiers and Independents. Clearly, other partitions should be tried as well--perhaps along socioeconomic characteristics. The party identification partition, nevertheless, should provide some minimal control over possible variations in $\underset{\sim}{A}$. Furthermore, since, across subsamples, we do not anticipate significant variations in r, in the number and relative ranking of the issues, and in relative candidate position, this model also provides a check on the model's sensitivity to purely stochastic differences.

Based on the eigenvalues of the adjusted covariance matrix, we decided that n = 2 was best for the Democrats and Republicans, whereas n = 3 was appropriate for the Independents. From a detailed analysis of the data it is clear that the Independents are a heterogenous group and that they exhibit the most perceptual variation. Therefore, we base the determination of the dimensionality of the candidate space for the Independents upon the results for the other two groups (see Cahoon, 1975 for details).

Table 2, however, shows the ordering of the candidates if we extract two dimensions. Although this ordering conforms in an intuitively satisfying way to a party identification or left-right breakdown, a negative axis weight is estimated for the Democrats. Virtually all variation in the candidate's positions occurs along a single dimension. Our hypothesis is that at least one valence issue is distorting these results.

Recall that there are sound theoretical reasons for supposing that if variations in the riskiness or familiarity of candidates is a salient concern, this concern should be given special treatment. Consequently, we hypothesize that respondents discount the thermometer scores of

TABLE 2. Ordering of Candidates on Principal Dimension with No Valence Issue

Republicans	Democrats
Johnson	Johnson
Humphrey	Humphrey
Muskie	Muskie
Kennedy	Kennedy
Romney	Rockefeller
Rockefeller	Nixon
Nixon	McCarthy
McCarthy	Romney
Agnew	Agnew
Reagan	Reagan

unfamiliar candidates. One problem with integrating this hypothesis into our analysis is the selection of an appropriate partition of the candidates. Here, we use some intuition and require, additionally, that all a's be positive and that the partition increase the across candidate correlation between the observed and predicted average thermometer score. While we cannot consider all possibilities, the partition that works best among those examined is: $\theta_{j3} = 0$ (Johnson, Nixon, Kennedy, Humphrey); $\theta_{j3} = 1$ (Muskie, Agnew, Reagan, Rockefeller, McCarthy, Romney).

The value of r which works best is ½, which is to say that respondents fail to discriminate sharply between candidates who are far from their ideal point. A review of the data reveals that the Democrats do not distinguish as well among the Republicans as they do among the candidates of their own party, and similarly for the Republicans. The analysis using r = 1 (straight distance) is almost as good, but squared distance is significantly poorer.

Using this assumption in our regressions to estimate $\underset{\sim}{\theta}$, $\underset{\sim}{A}$, the axis rotation, and $\underset{\sim}{\bar{x}}$, Figures 8, 9, and 10 show the resultant spatial maps (the symbol "0" denotes the estimated mean population preference, $\underset{\sim}{\bar{x}}$), while Table 3 provides several related statistics.

With respect to Table 3, the most important effect of including the valence issue assumption in our regressions is the increase in the relative salience of, and variability of the candidates over the vertical dimension. Aside from differences that we can attribute to rotation, the ordering of the candidates along the horizontal dimension conforms to Table 2. The vertical dimension, then, is the dimension that previously failed to discriminate candidates. Aside from this change, several additional aspects of these results warrant emphasis. First, note that aside from variations in distance, the candidates' relative positions are quite similar in all three figures. Second, the estimated mean ideal points conform to our expectations, i.e., the Republican mean preference

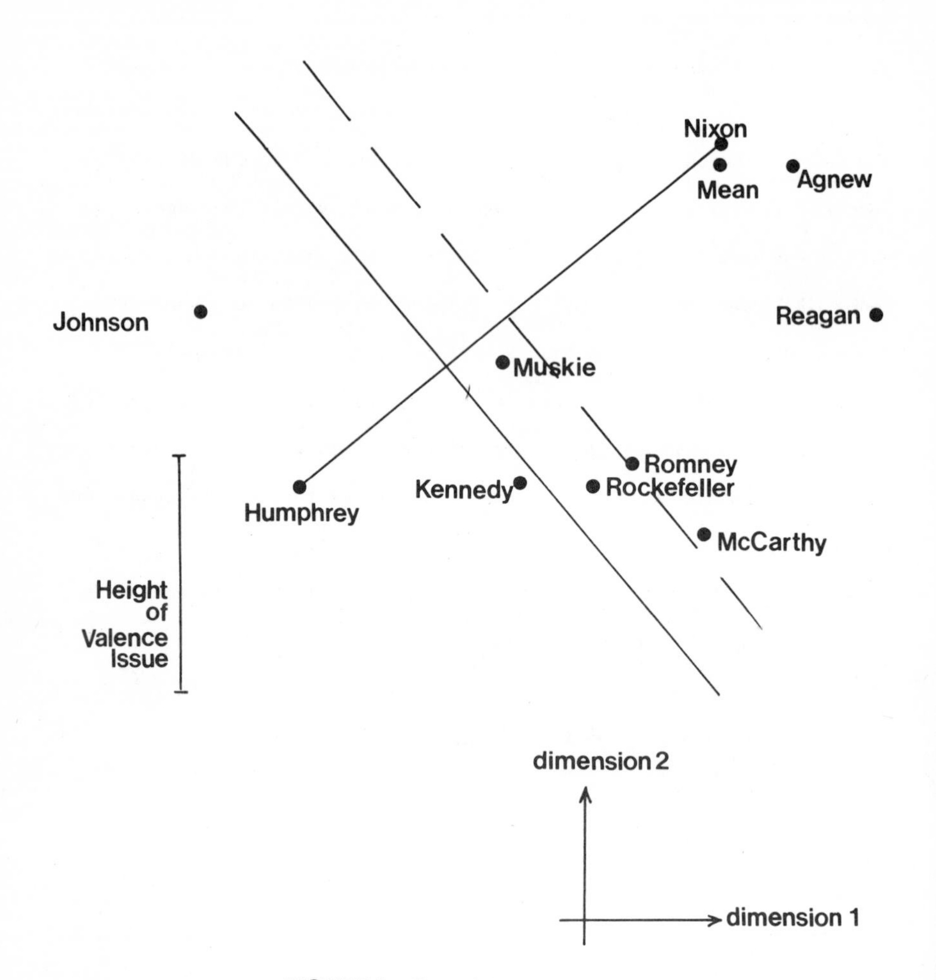

FIGURE 8: Republicans

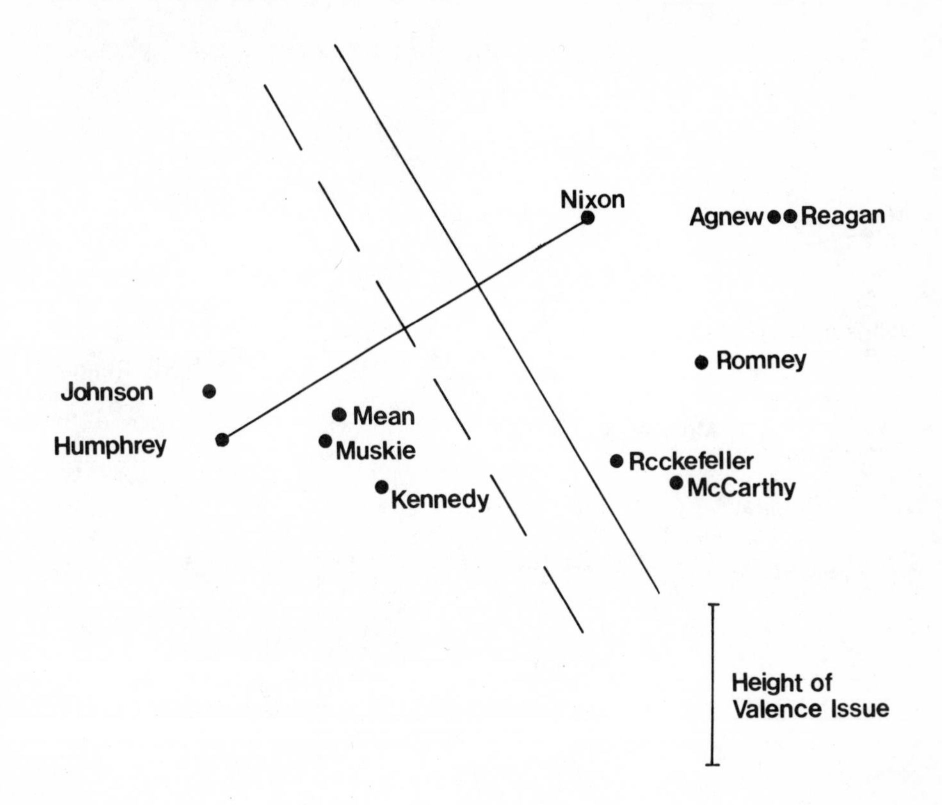

FIGURE 9: Democrats

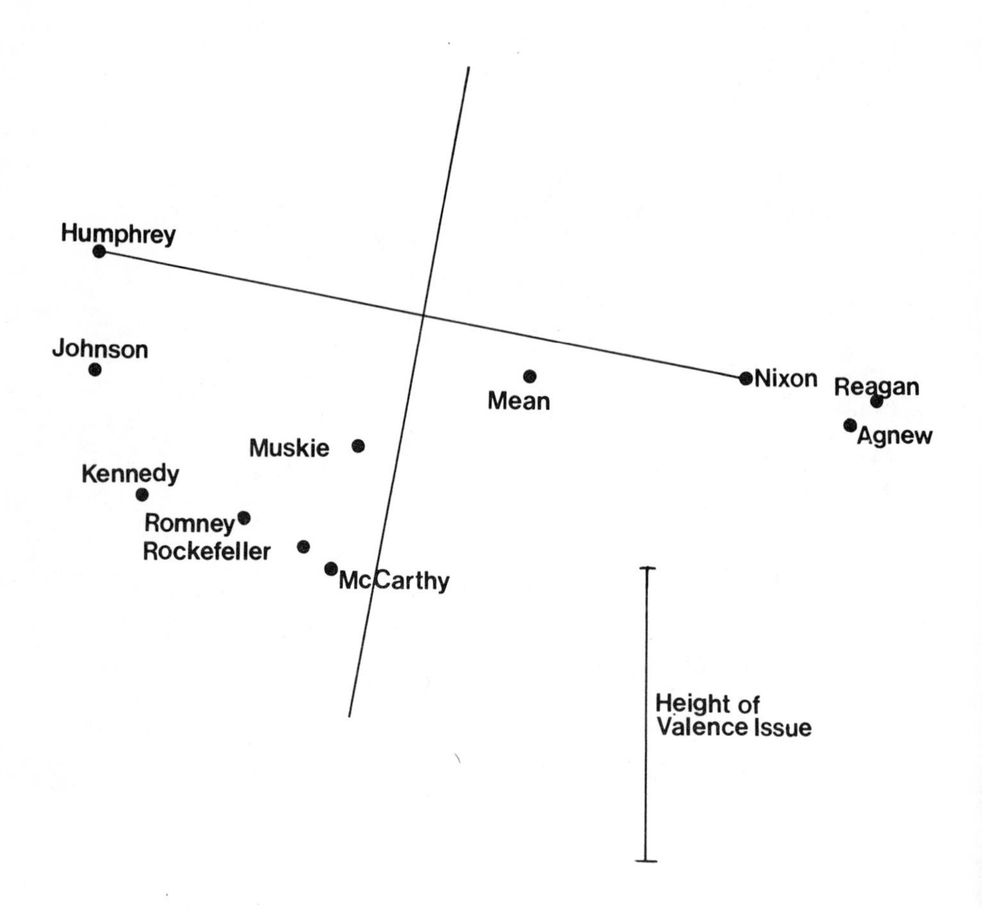

FIGURE 10: Independents

TABLE 3. Best Fit Predictions of the 1968 Vote and Related Statistics

	Republicans	Democrats	Independents
Predicted Vote			
Nixon	96.8	17.9	62.7
Humphrey	3.2	82.1	37.3
Actual Vote			
Nixon	92.6	21.6	62.7
Humphrey	7.4	78.4	37.3
% Vote Correctly Predicted	95.9	88.3	84.9
$\hat{a}_2/\hat{a}_1$	1.74	5.43	2.44
$\hat{a}_3/\hat{a}_1$	.79	.39	1.12
R^2	.995	.972	.983
r	½	½	½

is near Nixon, the Democratic mean is near Muskie and Humphrey, and the Independent mean lies midway between Muskie and Nixon. We note also that the inclusion of the valence dimension decreases the sum of squared errors in our regressions by a factor of two.

The one deviation from a consistent pattern across subpopulations concerns Independents. Specifically, while the order of the candidates along the vertical dimension for Democrats and Republicans places Nixon, Agnew and Reagan at one extreme, McCarthy and Kennedy at the other and Johnson, Humphrey and Muskie near the center, for Independents Humphrey is at one extreme, followed by Johnson and Nixon. Observe, however, that a 30° counter-clockwise rotation of the Independents' space brings that space in line with Figures 8 and 9. Given the quality of the data, a 30° rotational error seems well within tolerances--and is, in fact, consistent with the rotation we observed in Figure 7 using artificial data.

It is interesting at this point to compare the variance of candidate positions to the variance of citizen ideal points. The pattern is the same for all three groups and so, in Figure 11, we portray the distribution

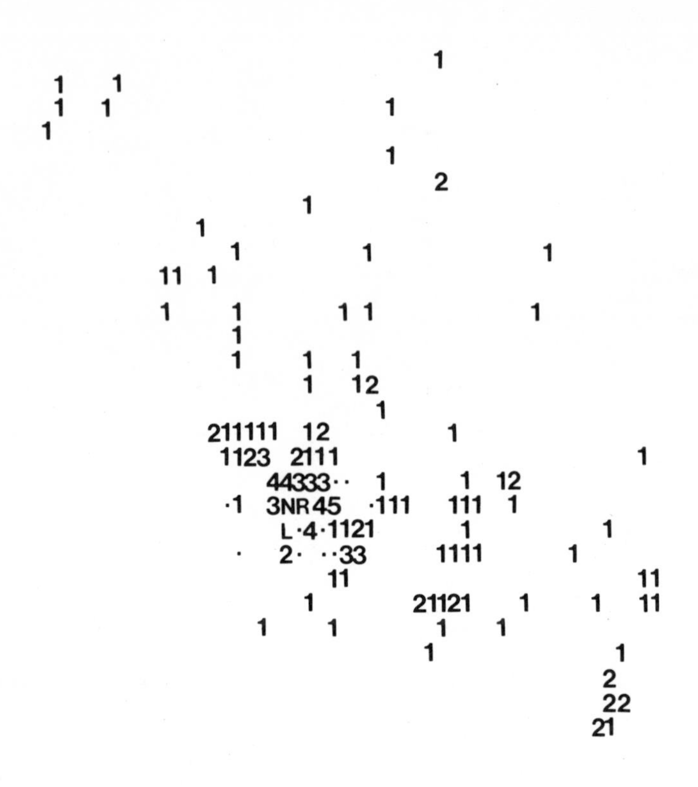

FIGURE 11: Distribution Of Republican Identifiers, With · Denoting Candidate Position

of Republican ideal points, letting "*'s" denote candidate position (numbers and letters are used to denote the number of ideal points at a particular point on the grid such that "1" = 1,..., "A" = 10, "B" = 11 and so on). Clearly, the dispersion of the $\underset{\sim}{x}_r$'s is far greater than of the $\underset{\sim}{\theta}_j$'s.

With respect to predicting votes, the dotted lines in Figures 8, 9, and 10 represent our theoretical predictor as to how citizens divide between Nixon and Humphrey. That is, all voters with ideal points to the left of this line are closer to Humphrey than to Nixon, and should, therefore, vote for Humphrey. All voters to the right of the line should vote for Nixon. The dark solid line corresponds to the line parallel to the bisector that maximizes the percent of the votes correctly predicted. We use this line in reporting the percentages of the votes for each candidate in Table 3. For Independents, the two lines are identical. For Republicans, the line is biased toward Nixon; and for Democrats, it is biased toward Humphrey. There are at least two explanations for this "bias."

The first explanation is statistical and does not relate to the model directly. However, it is seen most easily by utilizing the model. Consider, again, the predicted votes for the Democrats. If we predict votes based solely according to which candidate the citizen is closer, Nixon or Humphrey, we underestimate Humphrey's votes and overestimate Nixon's votes. This occurs because we utilize the estimated citizen positions to predict votes. Let us illustrate this effect by the following one-dimensional situation. Nixon and Humphrey are the only two candidates. We are considering the behavior of the Democrats, so we assume that the majority of the citizens are located nearer Humphrey. This situation is illustrated by Figure 12, where f(x) represents the true distribution of the ideal points in the population. We utilize $\underset{\sim}{\hat{x}}_i = \underset{\sim}{x}_i + \varepsilon_i$ to predict voting behavior, and the distribution of $\underset{\sim}{\hat{x}}_i$ is represented in the figure

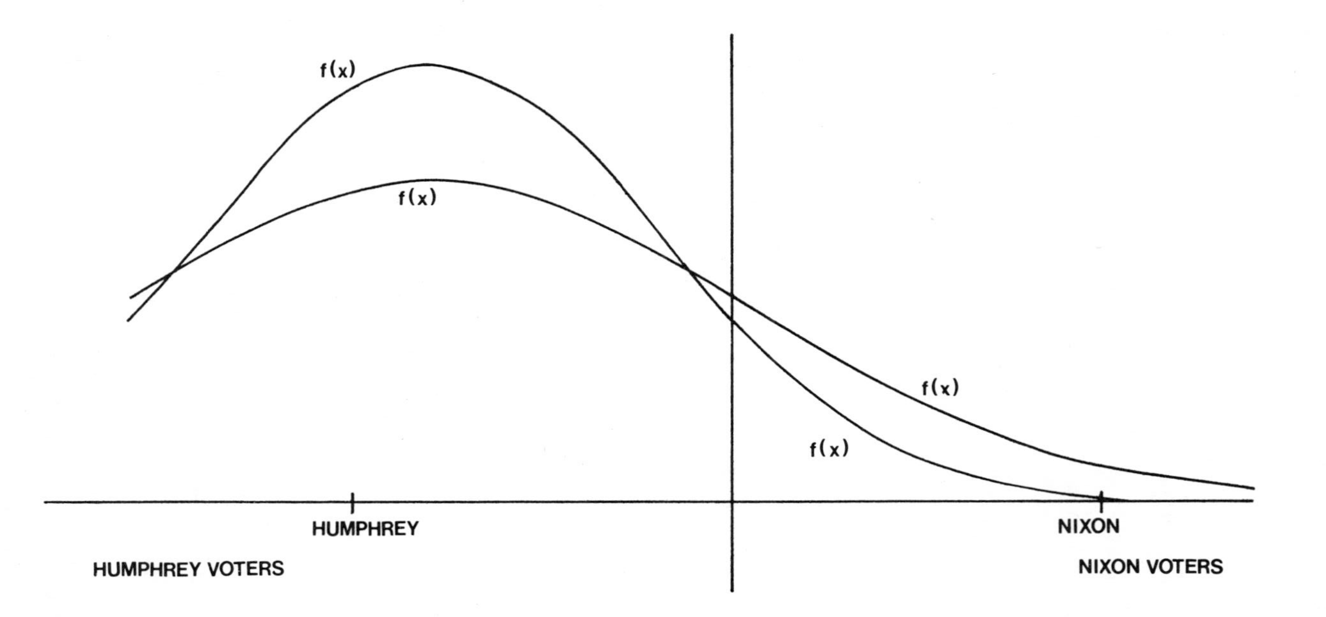

FIGURE 12 Probability Density Functions of the Ideal Points in the Population and the Estimated Ideal Points

by $f(\hat{x})$. The probability of voting for Nixon, then, is the area under $f(\hat{x})$ to the right of the vertical line. The probability of predicting a vote for Nixon based upon the estimated citizen position is the area under $f(x)$ to the right of the vertical line. Clearly, the probability of predicting a vote for Nixon is greater than the probability of a vote for Nixon. Thus, predicting votes based on the distance from the candidate, utilizing the estimated citizen positions, results in overestimating Nixon's vote and consequently underestimating Humphrey's vote. The effect is reversed if we predict votes for Republican respondents. Here we overestimate Humphrey's vote and underestimate Nixon's vote. There is no equivalent effect for the Independents, since the votes are more evenly distributed between Nixon and Humphrey.

This same reasoning shows that we increase the probability of correctly predicting votes if we move the line dividing the voters toward Nixon for the Democrats and away from Nixon for the Republicans. From Figures 8 and 9, it can be seen that this is exactly what has been done in order to correctly predict the maximum number of votes.

In addition to this reasoning, there is a second cause for bias. The thermometer responses were secured after the election, and we suspect that this affects our predictions. Note that for the Democrats, if we move Nixon closer to Agnew and Reagan, the disparity between the two lines diminishes. It is reasonable to suppose that Nixon's scores are enhanced by his new status as President-elect--he is moved closer to the center of the space than is representative of his correct position.

An important question, now, is whether these spatial maps are meaningful and useful. Aside from commenting that dimension 1 resembles a "left-right" continuum and that 2 principally differentiates by identification, our imaginations are not so limited doubtlessly as to preclude the construction of a good story about any observed dimension. There are other means, though, for assessing the substantive content of dimensions.

Included in the 1968 Election Survey are several questions that elicit from the respondent a thermometer score for several political categories, including "Viet-Nam War Protestors," "Liberals," "Republicans," "Democrats," and so on. Rusk and Weisberg use these scores to estimate (independent from candidates) the spatial configuration of these groups. They then overlay this space over the candidate space, to infer the substantive content of the original dimension. We illustrate here an alternative procedure: the scores for each group are used to divide respondents into five groups. Group 1 consists of all respondents who give a score from 0 to 19; Group 2 from 20 to 39; Group 3 from 40 to 59; Group 4 from 60 to 79; and Group 5 consists of respondents who gave scores from 80 to 100. Within each group we then calculate the mean of the estimated respondents' positions. This, then, yields five average positions that range from greatest dislike towards a category to greatest liking for that category. By plotting the positions of these five groups for each category we can then ascertain the relation between citizen preferences for candidates and issue relevant categories of political actors.

The thermometer issue question is applied to a total of nineteen issue-categories. Not all of these are useful, nor do all of them reveal any relationship to the respondents' positions. In Figure 13 through 17 we show some of the relationships between citizen feelings on the "issues" and their positions relative to the candidate. Table 5 presents the number of respondents in each of the five groups for each of these figures.

The relationship between Republicans and their feelings toward "Viet-Nam War Protestors" is shown in Figure 13. There is no thermometer question by which citizens could express their feelings on Viet-Nam directly, but the striking alignment of the groups on the vertical dimension indicates that the issue of Viet-Nam is related to that dimension.

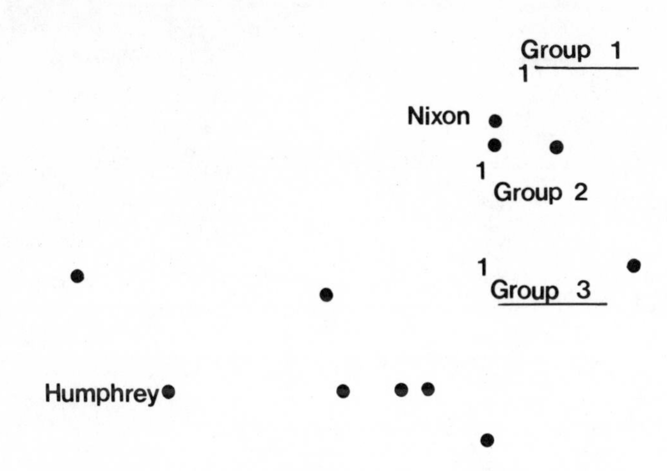

FIGURE 13

Republicans: "How Do You Feel About Viet-Nam War Protestors"*

* "Groups 4 and 5" are deleted since too few respondents are of this type.

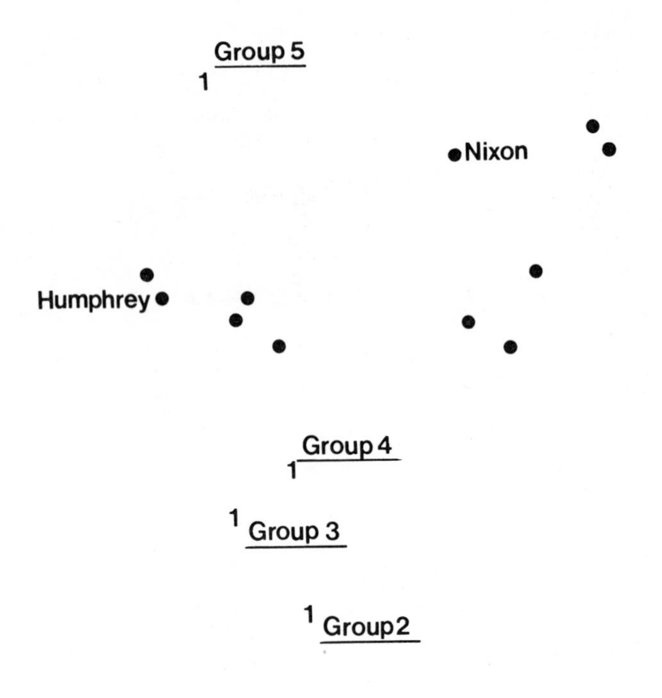

FIGURE 14

Democrats: "How Do You Feel About the Military"

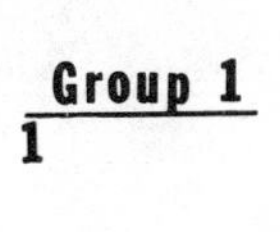

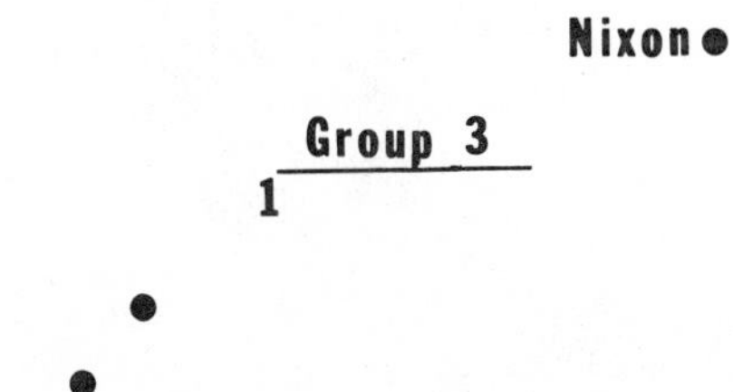

FIGURE 15

Democrats: "How Do You Feel About Liberals?"

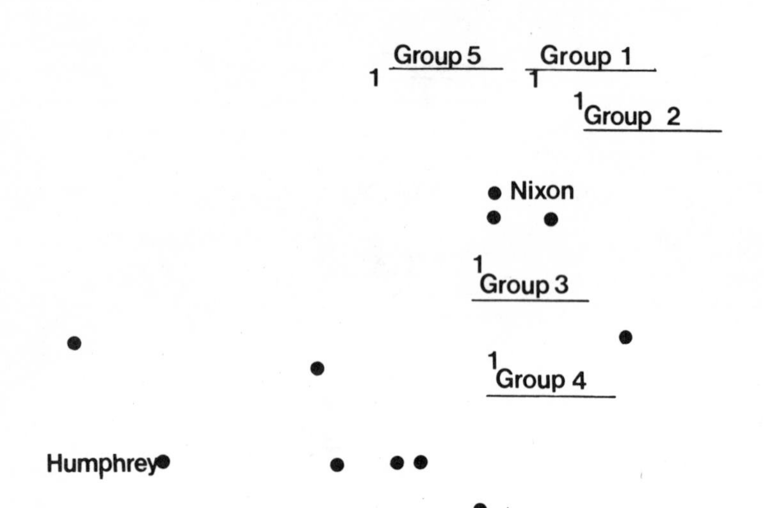

FIGURE 16

Republicans: How Do You Feel About Liberals

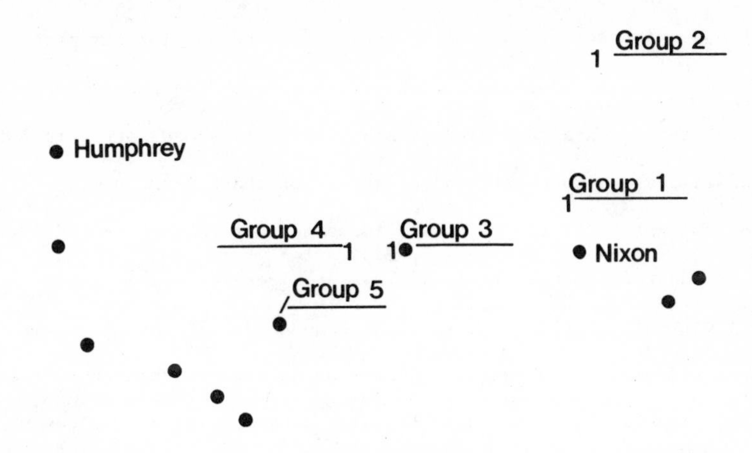

FIGURE 17

Independents: "How Do You Feel About Liberals?

The response of the Democrats to the same question is similar, but not as well defined. The Democrats express their attitude on Viet-Nam more clearly when asked to indicate how they feel toward "The Military." The plot of the respondents' positions on this issue is shown in Figure 14 and again we see a strong vertical alignment of the five groups.

The true nature of the vertical dimension is perhaps best indicated, though, by the respondents' feelings toward "Liberals." The plot of citizen preference groupings is shown in Figures 15, 16, and 17. Within the Democratic group, the vertical dimension is clearly related to this dimension. Thus we see that the vertical dimensions in fact represent a broad range of social issues that can be interpreted in "liberal-conservative" terms. The Republican response to this question (see Figure 16) is not as well defined. If we combine Groups 1 and 2, and combine Groups 3, 4, and 5, however, we see that Republicans as well view the vertical dimension in the same way as Democrats. The response of Independents (see Figure 17) reveals the effects of the rotation of their candidate space relative to the space we estimate for Democrats and Republicans. If we combine Groups 1 and 2, and combine Groups 3, 4, and 5, we see that the "liberal" alignment is rotated from the direction we observe for the Democrats and Republicans. This agrees with our earlier observation that the estimated candidate configuration for Independents disagrees with that of the Democrats and of the Republicans only up to a rotational error.

Other similar vertical alignments of groups occurs with respect to nearly all remaining thermometer questions. Of all the questions asked, only one differentiates citizens on the second dimension--party identification. Yet even here there is some correlation with citizen positions on the liberal-conservative dimension. Figure 18 indicates the feeling of Democrats towards "Republicans." Combining Groups 1, 2, and 3, and combining Groups 4 and 5, we see an alignment with the line connecting Nixon and Humphrey that intersects the vertical dimension at an approximately

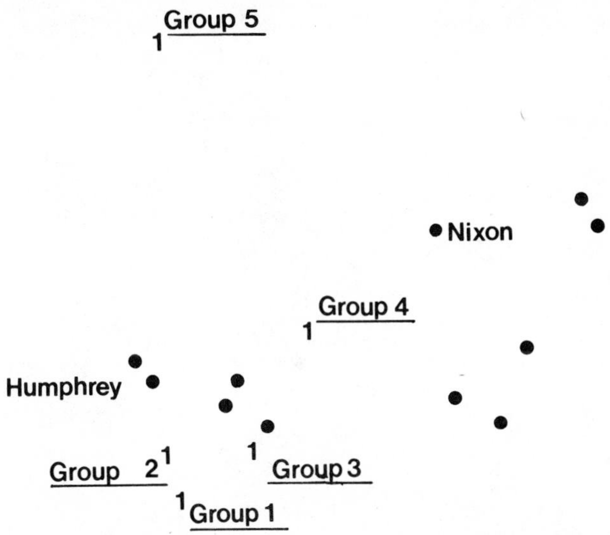

FIGURE 18

Democrats: "How Do You Feel About Republicans"

TABLE 4. Number of Citizens in Each Group in Following Figures

Figure	13	14	15	16	17	18	19
Party Group	Rep	Dem	Dem	Rep	Ind	Dem	Rep
Group 1: scores 0-19	117	11	17	34	23	21	25
Group 2: scores 20-39	36	11	18	26	14	14	15
Group 3: scores 40-60	74	74	79	137	112	185	135
Group 4: scores 61-80	4	58	49	30	27	50	43
Group 5: scores 81-100	9	161	51	14		45	23

Note: A score of 100 indicates a great liking for an "issue," while a score of 0 indicates a great dislike for the "issue."

45° angle. A similar pattern is observed in Figure 19 where we represent the feelings of Republicans for "Democrats." Here we can combine Groups 4 and 5 and combine Groups 2 and 3 to clarify the effect.

This, of course, does not exhaust the possibilities for "issue analysis," i.e., the mean by which we can attempt to secure a substantive interpretation of the recovered dimensions. Attitudes elicited on other questions in the survey can be used as well. For example, in a preliminary analysis of the 1972 Election Survey, we find that the thermometer scores on the inflation issue line up on the horizontal axis, which supports the conjecture that the party axis involves economic issues. We conclude that the dimensions in Figures 8, 9, and 10 "make sense" and reveal that citizens do not evaluate candidates simply in terms of party identification or in terms of ideology; instead, these two notions both operate (albeit in a correlated fashion with respect to the distribution of preferences) to form a citizen's evaluation of a candidate.

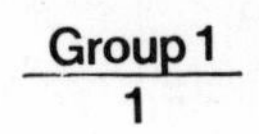

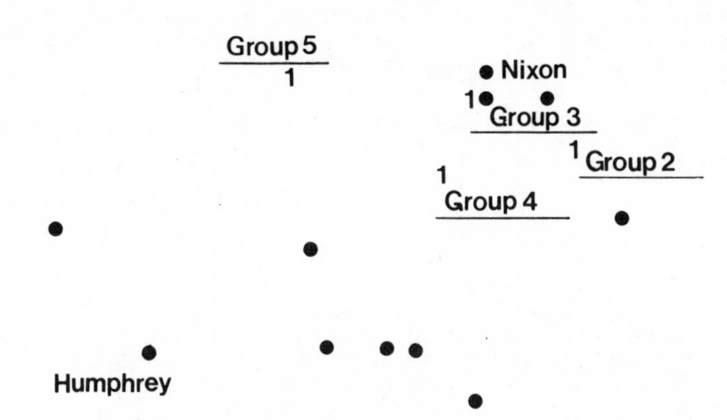

FIGURE 19

Republicans: "How Do You Feel About Democrats?"

REFERENCES

Cahoon, L. "Locating a Set of Points Using Range Only," Ph.D. Thesis, Statistics Dept. Carnegie-Mellon Univ. 1975.

Cahoon, L. and M. Hinich. "Locating Targets Using Range Only," IEEE Trans. on Information Theory IT-22 (2) (1976): 212-225.

Carroll, J. D. and J. Chang. "Analysis of Individual Differences in Multidimensional Scaling via an N-Way Generalization of 'Eckart-Young' Decomposition," Psychometrika 35 (1970): 283-319.

Davis, O. and M. Hinich. "A Mathematical Model of Policy Formation in a Democratic Society," In Mathematical Applications in Political Science, II, ed. by J. Bernd. Dallas: SMU Press, 1966.

Davis, O., M. Hinich, and P. Ordeshook. "An Expository Development of a Mathematical Model of the Electoral Process," American Political Science Review 64 (1970): 426-448.

Hinich, M. "Some Evidence on Non-Voting Models in the Spatial Theory of Electoral Competition," Forthcoming Public Choice (1978).

Rabinowitz, G. Spatial Models of Electoral Choice, Chapel-Hill: Univ. of North Carolina Social Science Monograph, 1974.

Rusk, J. and H. Weisberg. "Perceptions of Presidential Candidates," Midwest Journal of Political Science 16 (1972): 388-410.

Schoneman, P. "On Metric Multidimensional Unfolding," Psychometrika 35 (1970): 349-366.

Torgerson, W. Theory and Methods of Scaling, Chapter 11. N.Y.: J. Wiley, 1958.

A
B
C 8
D 9
E 0
F 1
G 2
H 3
I 4
J 5

3051411540